MW01705951

RAND

Army Forces for Operations Other Than War

Ronald E. Sortor

Prepared for the United States Army

Arroyo Center

Approved for public release; distribution unlimited

PREFACE

National military strategy has changed the focus of military planning to include a broader range of missions spanning the spectrum from major regional contingencies (MRCs) to operations other than war (OOTW). This change places additional demands on the Army that affect the required mix of active and reserve component forces. Planning for MRCs has presumed reliance on the active component for early-deploying combat forces and ready access to the reserve components for the bulk of support forces. However, for operations other than war—such as Somalia and Bosnia—the Army may not always be able to call on the reserve components for frequent or extended deployments. Nor may the active component be able to support OOTW missions while maintaining a ready MRC capability and meeting its other constraints. These conflicting demands lead to a key planning question: How should the active component and the reserve components be structured to meet the Army's evolving requirements?

This question was addressed in a RAND project entitled "Implications of Changing National Security Strategy for Army Active-Reserve Mix," sponsored by the Commanding General, U.S. Army Forces Command. The first portion of the research, documented in RAND report MR-545-A, *Army Active/Reserve Mix: Force Planning for Major Regional Contingencies,* focused on determining the forces required for major regional contingencies and on describing the Army forces planned for the late 1990s and early 21st century. This report describes a methodology for evaluating how OOTW missions affect the Army's ability to execute a major regional contingency with the planned force structure.

The analysis focuses on the effects of OOTW upon the active component, since it is the active component of the Army that has historically borne the burden of OOTW engagement and must also provide the units in the critical first days of an MRC while reserve units are being mobilized and prepared for deployment. The analysis uses data for operations conducted from 1975 through 1990 and for more recent operations in Somalia and Haiti. The methodology is applied in detail to the case of Somalia as an exemplar of an OOTW that the Army needed to perform while maintaining readiness for an MRC. It assumes that future operations will be conducted with a similar mix of forces and does not examine the potential for greater (or lesser) participation by Department of the Army civilians, contractors, or the other services. This analysis was largely completed during the very early days of the ongoing operations in Bosnia and thus does not capture lessons that may emerge from that experience.

The research was conducted in the Manpower and Training Program of RAND's Arroyo Center, a federally funded research and development center sponsored by the United States Army.

CONTENTS

FIGURES

TABLES

SUMMARY

National military strategy has changed the focus of military planning to include a broader range of missions spanning the spectrum from major regional contingencies (MRCs) to operations other than war (OOTW). These demands may conflict, leading to a key planning question: How should the active component and the reserve components be structured to meet the Army's evolving requirements? A previous Arroyo Center report dealt with the issue of forces for MRCs (Sortor, 1995). This report was prepared based on research addressing the second part of the mission mix—OOTW.

Operations other than war are not new to the Army. Over its entire history the Army has participated in operations such as disaster relief, humanitarian assistance, response to riots or insurrection, peace enforcement, military operations to restore order in foreign countries, refugee resettlement, and other limited military operations short of warfighting—operations referred to as OOTW. The Army has many capabilities that make it uniquely suited to respond to OOTW requirements both domestic and foreign. It thus finds itself called upon to perform a wide variety of tasks that are not part of its warfighting skill mix and that, in some cases, detract from its warfighting readiness.

Historically, OOTW requirements have been treated as a lesser included case in force structure planning and in equipping and training the force. Recently, however, two sets of events have occurred that bring into question whether OOTW demands can continue to be so treated. First, OOTW demands have increased in number and size, and there is a belief that they may increase further. Further, the character and circumstances of many of these operations have placed a greater burden on the Army to perform in the glare of television lights and in full view of the world. This has increased the need to be as efficient and effective as possible and to minimize any chance of either failing in the mission or performing it in a manner that causes national or international political embarrassment or invites condemnation. Second, as these demands have increased, the Army force structure has decreased significantly, and there are fewer units to share in the burden of OOTW. This makes it more difficult to maintain sufficient ready units for MRCs. These factors compel us to ask whether OOTW demands can continue to be treated as a lesser included case.

APPROACH AND METHODOLOGY

The research described in this report was directed at the question, How does the conduct of operations such as peace enforcement, humanitarian assistance, peacekeeping, and lesser regional contingencies influence the readiness and availability of Army forces to deploy to an MRC? We examine OOTW missions performed by the Army since 1975 and plans for possible future operations in order to define force requirements for OOTW. The requirements are defined in terms of unit type, numbers of units, and duration of deployment, and are intended to exemplify the forces likely to be required for future OOTW scenarios. We then compare the number, type, and component of Army units needed for simultaneous OOTW and MRCs to those in the planned force structure. The analysis does not examine the degree to which the operations might be conducted with a different mix of forces (i.e., active or reserve Army, Department of the Army civilians, contractors, or forces from the other services) in the future nor the potential effects of that different mix.

We also developed a methodology to evaluate the influence of OOTW on the Army's capability to deploy forces to an MRC, taking into consideration the time required for units first to prepare and deploy for the OOTW and then to regain combat-ready status after their return. The methodology also considers the effects of rotation and tour length policies on the availability of forces. That methodology is then applied in detail to the case of Somalia as an OOTW that the Army needed to perform while maintaining readiness for an MRC.

OOTW Factors Affecting Army Forces

The nonwarfighting operations that the Army has been called upon to perform have varied widely in size, duration, and type of forces required. This variation in type of operation, the size of force required, and the composition of the typical force for each is summarized graphically in Figure S.1. Typically, however, OOTW have not been large operations in the context of the total Army force structure.

RANDMR852-S.1

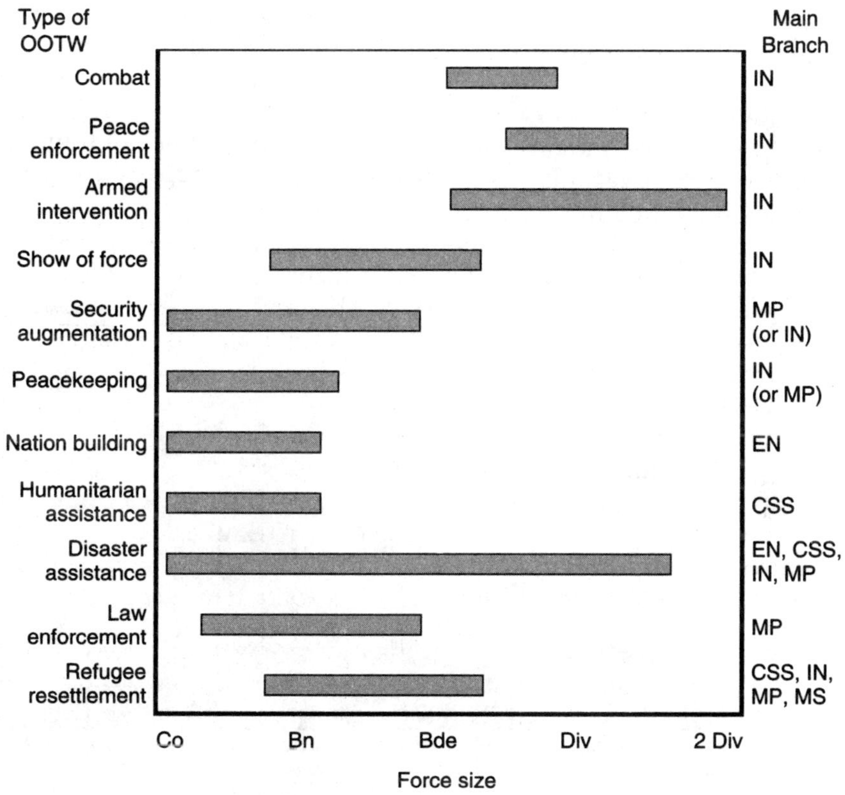

Figure S.1—Size and Composition of Operations Conducted Between 1975 and 1990 Vary Widely

In many operations, as in the case of Operation Restore Hope in Somalia, the total forces required and the mix of unit types have changed quickly during the course of the operation rather than in some more predictable and linear progression from force buildup, to the conduct of operations, to the subsequent withdrawal of forces. Figure S.2 shows the changing nature of the requirements in the case of Somalia and the shifting composition between combat, combat support (CS), and combat service support (CSS) as the needs of the operation changed over time.

Because OOTW often require skills and capabilities that are not emphasized in peacetime training for warfighting missions, units will spend time preparing for their OOTW tasks just before deployment. In the case of a fast-breaking contingency OOTW requiring immediate response, there may not be time for an extensive training period and units will primarily rely on their general soldier skills and discipline. In other cases, where there is time for a planning and training period, units may spend weeks or even months preparing for deployment.

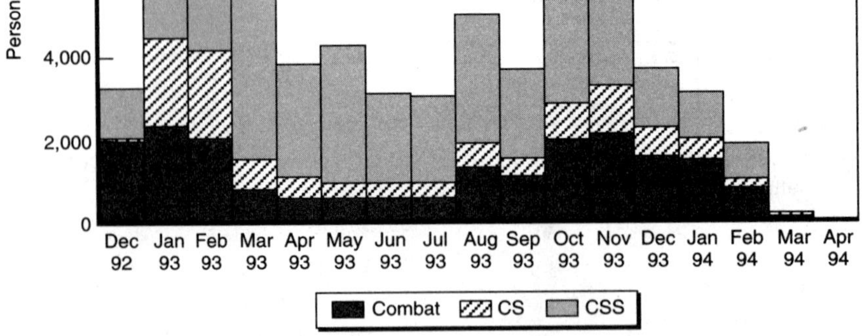

Figure S.2—Personnel Deployed in Somalia

Because OOTW tasks are different, to varying degrees, from warfighting tasks, the unit's readiness and capability for an MRC degrades over time as it is engaged in an OOTW. Upon return from an OOTW a unit must go through a recovery and train-up period to rebuild the skills essential to its warfighting tasks. This recovery period may last from only a few weeks to six months or more, depending upon the type of unit, the character of the OOTW, the length of the deployment, and the opportunity to engage in training on warfighting skills while deployed on the OOTW.

These preparation and recovery times become even more telling on the readiness of the force as units are rotated in and out of the OOTW. In the case of Somalia and Haiti, units were typically rotated about every 120 days. In addition to units deployed to the OOTW, some units are being trained for a subsequent rotation while others that returned from a previous rotation are regaining their warfighting skills and readiness for a possible MRC deployment. We do not believe most of these units would be able to deploy immediately to an MRC, although some units that had just begun preparing for deployment would still be sufficiently ready to go, as would many units that might be near the end of the desired recovery process (albeit at a reduced level of capability). In evaluating the effects of OOTW on Army forces, it is important to take into consideration these preparation and recovery processes as well as the effects of rotation and tour length policies.

ESTIMATING THE EFFECTS OF OOTW ON ARMY MRC CAPABILITY

Using the methodology discussed above, we estimated the effects on the Army's capability to deploy effective forces to an MRC while engaged in an OOTW like Somalia. The lower and darker portion of Figure S.3 depicts the total number of Army personnel (combat, CS, and CSS) deployed in Somalia for Operation Restore Hope between December 1992 and March 1994. That portion includes only the forces actually deployed in the theater, however, and therefore does not accurately portray the operation's full effects on the Army's capability for an MRC during this period. Combining the estimates for preparation, transit, and recovery time with the information on unit deployments for Somalia, we calculated the number of personnel in-

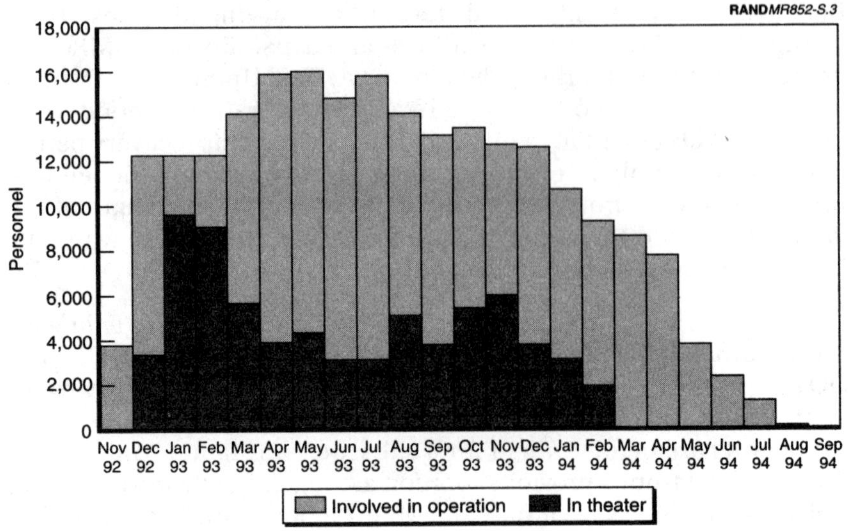

Figure S.3—Total Personnel Involved in Somalia Over Time

volved in the operation over time. The additional units engaged in preparation and recovery are depicted by the upper light-shaded portion of the bars in Figure S.3. Note that this increases the estimate of the personnel involved—and not available for an MRC—fourfold during much of the operation and indicates a significant number involved for four or five months after the majority of the forces have withdrawn from the theater.

Comparing the forces required for Somalia with those required for an MRC and the planned Army force structure indicates that for most types of units, the Army has an adequate number of units to both perform a limited number of OOTW and also maintain a ready MRC capability. That is not to say there are not some shortfalls, particularly in cases where the active component is engaged in an OOTW (as it was in Somalia) and where our analysis would indicate that reserve component units could not be ready in time to fill the early-deployment requirements for an MRC.[1] Figure S.4 shows the active com-

[1]See Sortor (1995) for the results of our analysis of MRC requirements and Army capabilities.

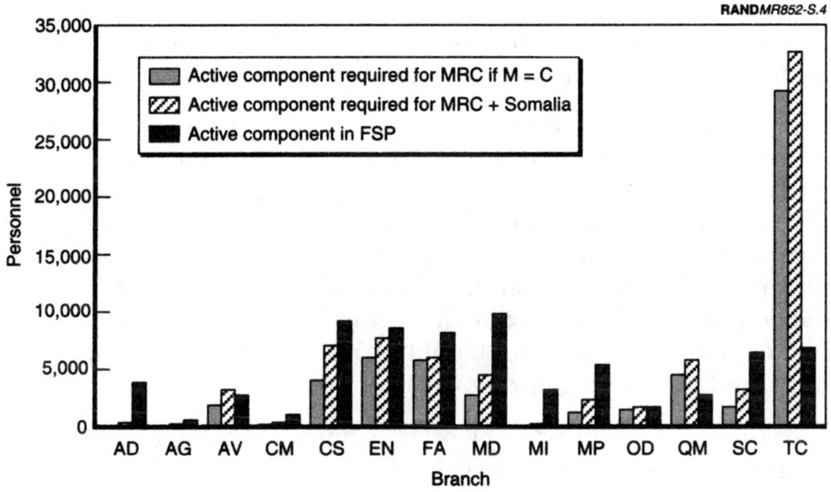

Figure S.4.—Comparison of Requirements and Active Component FSP Capability by Branch

ponent requirement for a Southwest Asia MRC scenario and the combined requirements for Somalia and the MRC, and compares them to the active component capability in the FSP (force support package).[2]

As shown in Figure S.4, there are shortfalls in capability (particularly in transportation and quartermaster) even in the absence of an OOTW. An operation like Somalia adds slightly to these shortfalls. These limited shortfalls, however, do not tell the whole story. The effects of operations like Somalia go beyond unit counts and rotation policies.

The Army often deploys only portions of units to an OOTW. A factor sometimes overlooked in assessing the impact of an OOTW is that if a significant portion of a unit is deployed, it is not just the deployed personnel who are unavailable, but also those left behind. Those left behind often do not have the right mix of personnel or the equip-

[2]The FSP consists of high-priority units (both active and reserve components) designated to meet the support force requirements for the first MRC.

ment either to effectively train or to deploy to an MRC until the deployed personnel and equipment return. This further increases the portion of the force whose readiness is affected by an OOTW.

Units other than those deployed to the OOTW theater are also affected because personnel are taken from them to augment the deploying units. In the case of Somalia, for example, the TPFDD (Time Phased Force Deployment Data) showed that ten military police companies were deployed, a total of 1,193 personnel. Personnel deployment data showed that the military police personnel (MOS CMF 95) who had deployed to Somalia came from 62 different units (41 military police companies, 10 military police battalion headquarters units, and 11 other organizations).

Cross-leveling of personnel often occurs because many of the support-type units needed in OOTW are not structured in peacetime with their full wartime-required complement of personnel authorized. Further, because of the drawdown and personnel assignment priorities, these units are also often undermanned relative to their peacetime authorized level. When called upon to deploy on a "real" mission, however, more often than not the unit must be brought to full wartime strength. Additionally, to execute an OOTW the Army does more special tailoring of forces than it does for MRCs, and this tailoring often departs from doctrine. Particular sections or platoons are required for a particular operation rather than the whole unit, or only a fraction of the capability is needed for the limited OOTW as compared to the warfighting task. This makes deployments to OOTW much more management intensive. These factors place additional burdens on all the units, whether deployed to the OOTW or not, and require an inordinate amount of command and leadership time and attention to be diverted from normal peacetime training and readiness activities in order to manage the preparation, deployment, sustainment, and recovery of units and parts of units deployed to OOTW.

Equipment readiness often suffers in OOTW because of the extra wear and tear caused by intensive usage in sometimes hostile environments and under circumstances that do not permit adequate routine maintenance and service. Support equipment is sometimes lost in transit or given away after the operation, and replacement of that equipment may take an extended period.

CONCLUSIONS

For most types of units, the Army has an adequate number of units to both perform a limited number of OOTW and also maintain a ready MRC capability. However, there are a few unit types in which the Army force structure is stressed. They are units typically needed both in OOTW and to meet early-deploying MRC requirements.

Support units like petroleum supply companies, water purification teams, maintenance teams, terminal operations teams and companies, and light-medium and medium truck companies are the ones typically deployed to an OOTW and in short supply in the active component. But shortfalls caused by OOTW requirements are not large in number. The cost to completely offset the OOTW effects would not be large in terms of the active endstrength required to field the additional active units.

Additionally, for some types of units that are in short supply, there are alternatives to buying more active force structure. For example, cargo documentation teams are required in the theater for an MRC by about day fifteen (by day seven for prepositioned ships). Only three automated cargo documentation units exist in the active component, and two were used in Somalia. Our previous analysis indicated that reserve component units of this type could confidently be relied on to arrive in the theater by about day eighteen. If a possible two- or three-day delay is acceptable, then the current force would be adequate. If not, with only a few additional resources, RC cargo documentation teams could be made ready more quickly and replace active component units engaged in an OOTW and therefore not immediately available for the MRC. Similarly, it is likely that other reserve component units like water purification teams, terminal operations teams, and perhaps larger units like truck and petroleum supply companies could also be made available more quickly. These units tend to be single-function support units that call for civilian-type skills and are the type of units the reserve components have demonstrated they are best at providing. Additionally, some of these capabilities are available from the private sector through contract support agreements.[3]

[3]For example, contract services provided under contract from Brown and Root have supported operations in Somalia, Haiti, and more recently Bosnia.

Future Role of Reserve Components in OOTW

Like the active component of the Army, the reserve components also have a long tradition of OOTW involvement, particularly in SOUTHCOM. They routinely participate in nation building, drug interdiction, and border control operations. However, for many OOTW contingencies, the circumstances of the operation and the nature of citizen-soldier service limit greater direct participation. Nevertheless, as discussed above in the context of cargo documentation teams, there are options and alternatives for the reserve components to play a greater role in mitigating the effects of the unpredictable nature of OOTW and the day-to-day status of units (both active and reserve).

As mentioned earlier, the Army often deploys only portions of units to OOTW. Adding a section or platoon to an active component unit manned by part-time reservists would provide the additional capability needed to offset the effects of limited OOTW deployment and permit the early deployment of a fully manned TOE unit in the event of an MRC. In other units it might be necessary to place authorizations for reserve soldiers throughout the unit rather than in sections and platoon formations.

When whole units are typically deployed to an OOTW, it may be better to place greater emphasis on making reserve units more ready so that they might deploy earlier in the event of an MRC. This does not mean keeping some or all of the reserve units more ready all the time. It is likely to be much more cost-effective to establish a process and procedures for real-time management of unit and force readiness based on current force needs, problems, and commitments. This would enable the deployment of reserve component units to an MRC earlier than planned when active component units are engaged or committed to engage in an OOTW, without the expenditure of resources needed to constantly maintain a higher state of readiness even when it is not required by the current situation.

Clearly, if in the future one or more large OOTW involved two Army division equivalents, as envisioned in the BUR, then virtually all of the active component support force would be required, with nothing left in the event of an MRC. This would require immediate presidential selected reserve call-up (PSRC) or mobilization so that reserve component units could either be deployed to the OOTW or placed on active duty for training so they could be made ready for early deployment in place of active component units in the event of an MRC. And of course, if future force reductions eliminate active component support units without offsetting increases in the readiness of reserve component units, the result would place additional stress on the Army's ability to execute OOTW and maintain ready MRC forces.

Summary

In summary, our analysis indicates that for the most part the present force is adequate in unit type and number, and OOTW requirements add only very slightly to some of the shortfalls in the Army-desired MRC capability—shortfalls that already exist in the absence of an OOTW. It has also been noted that while OOTW requirements do place somewhat different demands on the Army than do the warfighting operations, many allow the substitution of capabilities, particularly in the more stable types of operations where the threat is less demanding; thus the effects of OOTW can be further mitigated.

However, our analysis does highlight the need to consider OOTW effects beyond the units actually deployed to an operation. Cross-leveling, tailoring, and deployment of partial units all place added demands on the Army's ability to manage the readiness and availability of the force. These OOTW demands must be accommodated by changes in unit structure and manning for the Army to have a sufficiently robust capability to meet likely OOTW and planned MRC demands in an acceptable manner. The reserve components can play a greater role in this regard, though probably not through increased direct participation in OOTW contingencies.

ACKNOWLEDGMENTS

This report benefited greatly from the support and assistance provided by many people in the Army and at RAND. It could not have been prepared without the considerable input and assistance from the staff at U.S. Army Forces Command Headquarters. I owe a special thanks to David Pearson who, while on a one-year exchange tour at RAND from the UK Ministry of Defence, developed much of the deployment data for Somalia and Haiti and the models used in the unit and force availability analysis. I also owe particular thanks to Patricia Dey for her work in accessing the various data and building the files and data analysis tools used in the analysis. At RAND, I received very helpful comments and reviews from Dave Kassing and Dave Persselin. The content and conclusions, however, remain the sole responsibility of the author.

GLOSSARY

AC	Active Component
ACR	Armored Cavalry Regiment
AD	Air Defense
AG	Adjutant General
ALO	Authorized Level of Organization
APOE	Aerial Port of Embarkation
AR	Armor
ARNG	Army National Guard
AT	Annual Training
AV	Aviation
BUR	Bottom-Up Review
CA	Civil Affairs
CAA	Concepts Analysis Agency
C-day	Date that deployments begin
CFP	Contingency Force Pool
CINCCENT	Commander in Chief, Central Command
CM	Chemical
CONUS	Continental United States

CS	Combat Support
CSS	Combat Service Support
DAIG	Department of the Army Inspector General
EAC	Echelons Above Corps
EAD	Echelons Above Division
EN	Engineer
FA	Field Artillery
FASTALS	Force Analysis Simulation of Theater Administrative and Logistic Support
FI	Finance
FORSCOM	Forces Command
FSP	Force Support Package
IN	Infantry
IRR	Individual Ready Reserve
JA	Judge Advocate
LAD	Latest Arrival Date
LG	Logistics
MAINT	Maintenance
M-day	Mobilization day
MD	Medical
MH	Military History
MI	Military Intelligence
MOOTW	Military Operations Other Than War
MP	Military Police
MRC	Major Regional Contingency
MS	Medical Services
NEA	Northeast Asia

OCONUS	Outside CONUS
OD	Ordnance
ODS	Operation Desert Shield/Storm
OOTW	Operations Other Than War
PA	Public Affairs
POD	Port of Debarkation
PSRC	Presidential Selected Reserve Call-up
QM	Quartermaster
RC	Reserve Component
RLD	Ready-to-Load Date
SC	Signal
SPOE	Sea Port of Embarkation
SWA	Southwest Asia
TA	Theater Army
TAA	Total Army Analysis
TC	Transportation
TDA	Table of Distribution and Allowance
TF	Task Force
TOE	Table of Organization and Equipment
TPFDD	Time Phased Force Deployment Data
TTAD	Temporary Tour of Active Duty
TTHS	Trainees, Transients, Holdees, and Students
UIC	Unit Identification Code
USAR	U.S. Army Reserve

Chapter One

INTRODUCTION

The national military strategy and the U.S. Army force structure have both been undergoing rapid and large change since the late 1980s. During the Cold War, the planned use of force and the consequent force structure sizing were tied clearly to deterrence of major conflicts where the most fundamental national interest might be threatened, i.e., survival of the nation and its institutions. The post–Cold War era presents a more complex planning problem both in terms of when and where to plan on the use of force and in terms of the consequent military force required. Vital interests may no longer provide an adequate framework for developing military requirements.

Planning for major regional contingencies (MRCs) provides a somewhat familiar framework, but a different set of challenges is presented by other demands for military forces—for operations such as Joint Endeavor in Bosnia, Restore Hope in Somalia, and Provide Comfort/Southern Watch in the Persian Gulf. These types of operations, termed operations other than war (OOTW), are considered by many to present the most likely and most frequent demand for U.S. military forces in the near future. Recent history would tend to support that view. Concern has been expressed that these operations may place such a burden on the U.S. military as to render it incapable of responding to major regional contingencies where our national interests are seriously threatened. The research documented in this report was undertaken to explore these concerns and determine the potential effects of OOTW on Army capability to respond to MRCs.

CHANGING STRATEGY AND FORCE STRUCTURE

Since the end of the Cold War, U.S. military forces have been reduced by substantial amounts. The number of personnel in the active Army, for instance, declined from 732,000 in 1990 to about 495,000 in 1996, a reduction of one-third. These reductions were driven, in part, by two successive reviews of national military strategy: the Base Force review during the Bush administration and the Bottom-Up Review (BUR) during the Clinton administration.[1] Both reviews recognized the need for military readiness to respond to two MRCs that could occur at the same time (for example, conflicts in the Persian Gulf and in Northeast Asia); and neither explicitly considered the force structure requirements imposed by OOTW, such as the recent operation in Bosnia. For this analysis we took as the baseline the present military strategy and planned forces that resulted from the BUR.

The first phase of this research focused on the adequacy of the evolving Army force structure, both active and reserve, to carry out Army missions and responsibilities associated with MRCs (Sortor, 1995). The analysis, discussed further in the next chapter, showed that under the current DoD planning assumption that four to five divisions are adequate for each major regional contingency, the BUR combat force is more than sufficient even when judged against a scenario with two nearly simultaneous contingencies of this size.

In contrast to the case for the combat forces, it did not appear that the planned *support* force structure would be adequate to provide the number of units at the needed readiness level to support two nearly simultaneous contingencies, even if the contingencies required only four or five combat divisions each. The pool of high-priority support units would be adequate for one modest-sized contingency but not for a single large contingency or two nearly

---

[1]As this report was being finalized in early 1997, the Department of Defense was preparing for a third such exercise, entitled the Quadrennial Defense Review (QDR). The QDR is examining both national military strategy for the future and the forces needed to carry out that strategy. Two aspects of the QDR debate have revolved around questions regarding resource limitations and the need for force structure versus modernization programs and the extent to which OOTW should be considered in determining the recommended force structure. These questions are likely to continue for some time.

simultaneous contingencies. The other support units in the planned force structure, given their lower priority for resources, would not be ready to deploy in time.

As demanding as the requirements for the MRCs may be, the Army does not have the luxury of focusing all of its attention and peacetime resources on preparing for them. It must also meet its other peacetime responsibilities, such as keeping forces stationed overseas to maintain forward presence and to prepare for other missions it may be given, such as peacekeeping and humanitarian assistance. For example, on August 5, 1994, the Army, in addition to the forces permanently stationed overseas, had 15,941 soldiers performing 1,826 missions in 105 countries (HqDA, 1994a). These deployments included an infantry battalion in the Sinai as part of the Multi-National Force and Observers (MFO), medical personnel supporting a hospital in Moldova, an air defense battalion reinforcing Korea, and 573 soldiers providing humanitarian assistance in Rwanda. With the deployment to Bosnia for Operation Joint Endeavor, the number of soldiers deployed overseas had grown to over 27,000 on April 18, 1996 (*Army Times,* July 1, 1996).

While operations other than war are included in the missions the military may be called upon to perform, forces are not sized based on OOTW mission requirements.[2] The assumption in the BUR was that forces engaged in OOTW would be withdrawn in the event they were needed for an MRC. There has been a general concern that this may not be possible or that it may take too long to withdraw the forces and prepare them for redeployment and employment in an MRC. Planning for MRCs has presumed reliance on the active component for early-deploying combat forces and ready access to the reserve components for the bulk of all support forces and, should they be needed, for later-deploying combat forces. However, for some OOTW the Army may not be able to call on the reserves for frequent or extended deployments. Nor may the active component be able to

---

[2]Missions included in military operations other than war (MOOTW) by Army FM-100-5 are peacekeeping operations, peace enforcement operations, humanitarian assistance operations, support to counterdrug operations, noncombatant evacuation operations, counterterrorism operations, counterinsurgency operations, show of force, and arms control. OOTW also includes nonmilitary missions such as disaster relief and nation building/assistance.

support these missions and at the same time maintain a ready MRC capability and meet its other peacetime commitments and constraints. The Army has stated for some time that the increasing pace of these peacetime operations threatens its ability to meet the requirements of the two-MRC scenario (HqDA, 1994b). These conflicting demands lead to a key planning question: How should the active and reserve components be structured to meet these evolving Army requirements?

OBJECTIVES OF THIS STUDY

The overall objective of our research was to address the question just stated. The previously reported work examined the implications of major regional contingencies for both combat and support forces with an eye to determining which could come from the reserve components and which must be in the active component (Sortor, 1995). In making that determination, we considered not only force structure but also the likely readiness levels of the reserve component units and their availability. We did so, however, without consideration for OOTW commitments. The research documented in this report examined OOTW from a historical perspective, the Army force requirements for OOTW, and how meeting these requirements might affect the Army's readiness to engage in MRCs. This phase of the research effort also addressed whether alternative active and reserve structures and relationships would better serve to meet the evolving requirements.

APPROACH

Our approach, in this phase of the research, was threefold: define force requirements for OOTW, investigate factors associated with OOTW that affect unit availability for an MRC, and develop a model describing the relationship between unit availability and unit rotation policy and how these factors interact to affect force availability. First, we define the force requirements by examining past and current OOTW missions, the types and numbers of units required, and the duration of deployment. Second, taking into consideration the force requirements, the time required for units to prepare, transit, and deploy for an OOTW, and the time required to regain MRC ready status, we evaluate the influence of OOTW on a unit's capability to deploy to an MRC. Third, we estimate the effect of OOTW on force availability by examining the interaction between the number of units required for the OOTW, the unit rotation policy, and the length of time units require to return to readiness.

In this analysis we assume that the capability to respond to a single MRC must be maintained while simultaneously conducting OOTW. Therefore, sufficient combat and support forces must always be available, at sufficient readiness, to respond to a single MRC. We will assume that forces engaged in an OOTW, while not available for the first MRC, would be available for any second MRC that might arise while forces were still engaged in the first.

ORGANIZATION OF THIS REPORT

The remainder of this report focuses on understanding how OOTW affect the Army's capacity to meet its MRC missions. Chapter Two provides the background for our analysis and summarizes our previous work on requirements for MRCs; it also briefly describes the Army requirements implied by the current military strategy, the missions identified to carry out that strategy, and the planned Army force structure. It describes the range and characteristics of operations we include under the OOTW umbrella. Chapter Three examines Army requirements for OOTW in terms of their characteristics, forces required, and frequency. Chapter Four focuses on the demands placed on units as a result of OOTW and how these demands could impact a unit's readiness for MRC deployment. In Chapter Five we investigate whether large OOTW deployments from limited CONUS-based forces could limit the Army's ability to react quickly to an MRC, and if the impact could be more severe if the intervention extended over a period in which fresh units would need to be rotated into the theater. Chapter Six discusses the role of the reserve components in OOTW, and Chapter Seven presents the conclusions from our analysis.

Chapter Two

BACKGROUND

Two watershed events—the end of the Cold War and the overwhelming success of Operation Desert Storm (ODS)—continue to shape the evolving national military strategy and the composition and size of the military forces needed to carry out that strategy. Another major influence is the continued demand for reductions in defense spending. One of the central public policy issues of the post–Cold War era is the level of resources for national defense and how much is likely to be affordable in the future. Today's forces and the current planned forces are the result of two previous reviews of national military strategy completed since 1990.

The end of the Cold War prompted immediate calls for decreasing the U.S. spending on military forces and the diversion of the "peace dividend" to other competing priorities. The first review of national military strategy and the changing military requirements after the fall of the Berlin Wall was premised on the continued decline of the Soviet Union and resulted in the Base Force proposal by the Bush administration. The results of that review came under almost immediate challenge for being too conservative, and many called for further reductions to free federal budget resources for other purposes.

Representative Les Aspin was one of the more vocal critics of the Base Force plan, and he proposed four alternative force structures, all of which called for a smaller U.S. military. Upon becoming Secretary of Defense in the Clinton administration, Aspin led the second review of post–Cold War military strategy and force requirements, the Bottom-Up Review (BUR), completed in September 1993. It has formed the basis for current force and resource planning (DoD, 1993).

The Bottom-Up Review identified four new dangers that would shape future military strategy and force structure needs:

- proliferation of weapons of mass destruction and delivery systems,
- regional dangers,
- danger to democracy, reform, and civil order, and
- danger of a weak economy.

The Bottom-Up Review concluded that conventional forces should be oriented primarily to address the regional dangers and the danger to democracy, reform, and civil order. It sized those forces based on three requirements:

- major regional contingencies,
- peace enforcement and intervention, and
- forward presence.

As a result of the BUR, the Army force structure was to decrease from the twenty-eight divisions (eighteen active and ten reserve) that existed in 1990 to eighteen divisions (ten active and eight reserve). In terms of military personnel, the Army would go from 1,486,000 (732,000 active and 754,000 reserve) to 1,070,000 (495,000 active and 575,000 reserve).

Table 2.1 shows the Army forces, by major force elements, used in our analysis.

Figure 2.1 depicts the Army force structure in the context of the Army's force generation model. This model arrays the forces according to their role in contingency operations and the time sequence in which the units would be needed to respond to a major conflict. The Major Contingency Response Force would be needed immediately to respond to the first MRC. If the MRC was in the Pacific theater, the forces would reinforce the Rapid Regional Response Force located in the Pacific theater. In terms of the support forces, FSP (force support package) 1 supports the EAD CS/CSS requirements for the lead

Table 2.1

Army Endstrength by Major Force Elements

Elements	Active Component	Reserve Components	Total Army
TOE units	311,300	487,900	799,200
Combat and special operations forces	184,900	185,300	370,200
Echelons above division/echelons above corps (EAD/EAC) support	126,400	302,600	429,000
Table of distribution and allowance (TDA)	123,900	87,100	211,000
Trainees, transients, holdees and students (TTHS)	59,800	0	59,800
Total	495,000	575,000	1,070,000

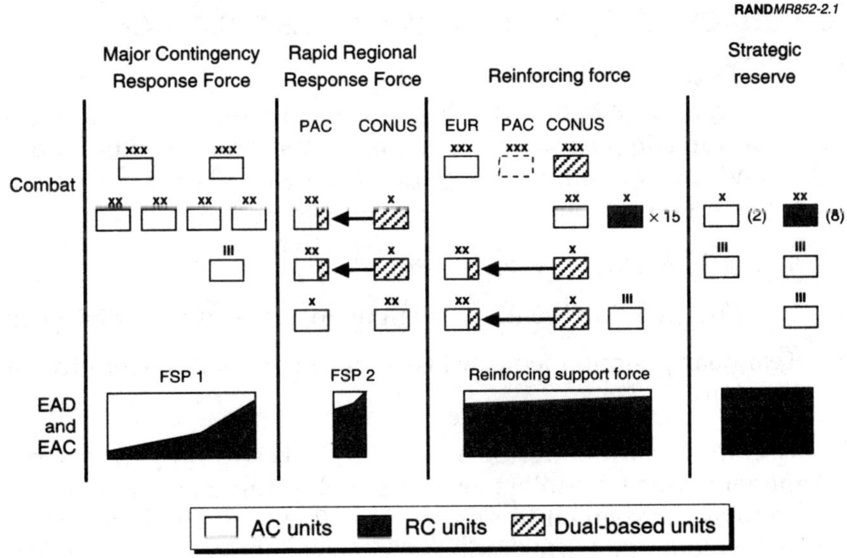

Figure 2.1—Army Force Generation

corps, with four divisions and one ACR. FSP 2 supports EAD CS/CSS for the remaining division, the EAC CS/CSS for the theater, and the planning headquarters for a second corps. In the remainder of this report we will include forces from both FSP 1 and 2 and refer to them as FSP. Forces from the reinforcing force would be available to reinforce the first MRC or to deploy to a second MRC if required. Additional forces are available in the strategic reserve, but at a lower state of readiness, and they would require substantially more time to prepare after mobilization before deployment to a major conflict.

While these are the forces that are currently planned, and therefore used as the baseline in our analysis, it is recognized that they may change in the future. Indicative of continuing pressures to find defense savings, in 1996 OSD suggested during budget deliberations that the Army look at further reducing its active structure to 475,000. With tight resource constraints, all of the services face difficult tradeoffs between support for current forces and modernization programs for future forces.

ADEQUACY OF PLANNED ARMY FORCES FOR MRCs

For the convenience of the reader, this section summarizes our previous analysis of MRCs (Sortor, 1995). That analysis focused exclusively on the adequacy of the evolving Army force structure, both active and reserve, to meet timetables for MRCs. Three cases were examined in some detail:

- Two separate MRCs, each requiring four to five divisions;
- One MRC requiring reinforcement up to a total of eight divisions;
- Two nearly simultaneous MRCs requiring a total of eight to ten divisions.

Using empirical estimates of how long it takes to prepare reserve component units for deployment after mobilization and a force mix methodology developed in our previous RAND research, we compared the requirements for various regional contingencies to the mix and composition of the proposed Army force.

Our analysis showed that under the DoD assumption that only four to five divisions would be needed for each major regional contingency, the Army combat force, shown in Table 2.1, would be adequate even when judged against a scenario with two nearly simultaneous contingencies of this size. However, the planned support force structure would not provide the number of units at the needed readiness level to support these same contingencies.

For a long time the Army has been heavily dependent on the reserve components for combat support and combat service support units at echelons above division to support active Army combat forces in all but the smallest of combat operations. In ODS, the first call-up of reserve forces in more than two decades, the Army deployed some 70,000 guardsmen and reservists in this role with good results. Since ODS, however, the Army has reduced the number of units and personnel and made changes in the way reserve component units are aligned and in the resourcing priorities.

We found that for a single Southwest Asia MRC requiring no more than four to five divisions and where mobilization occurs at C-day, the active component would be required to provide about 37,000 of the total 180,000-person support force requirement, and the reserve component could provide the remaining 143,000.[1] The planned mix of active and reserve component units in the FSP total about 151,200 personnel (70,600 active and 78,400 reserve) and would not be able to meet this requirement without some degree of risk. Further, as will be discussed later, a few branches would suffer substantial shortfalls.

As depicted in Figure 2.2, if mobilization were delayed for twenty days and occurred at C + 20, as was the case in the Persian Gulf War, the reserve component can be expected to meet only about 70,000 of the requirement, with the active component needing to provide 110,000 of the support personnel. This is far more support forces than would exist in the active component CONUS-based force under current plans.

---

[1]The term "mobilization" is normally associated with a declaration of national emergency and partial or full mobilization. The day mobilization occurs is termed M-day. The president also has authority, without declaring a national emergency, to order up to 200,000 selected reservists to active duty under his presidential selected reserve call-up (PSRC) authority. The day the call-up occurs is termed S-day. We will use M-day in this report to mean either presidential call-up (S-day) or mobilization (M-day).

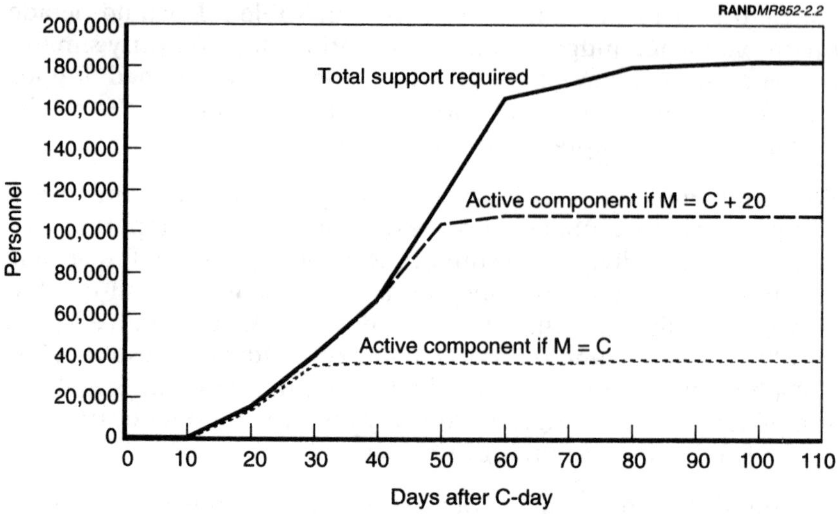

Figure 2.2—Mix of Active and Reserve Support Forces

For the planned force to provide the required support structure in terms of size and mix for a single MRC requiring four to five Army divisions, reserve component support units must be made available at or very soon after initial deployment of combat forces. In addition, Army planning scenarios often call for the forces to deploy more quickly than in ODS or in the scenario used in the above Southwest Asia example. A faster deployment of forces in the very early phase of operations places greater demands on the active component forces, since reserve component forces cannot be ready in time. In TAA01, for example, the Army planned on the basis of the support forces closing about ten days faster than in the illustration discussed above.[2] The effect is to increase the total active component requirement from about 37,000 to 59,000 even if mobilization is declared immediately. Further, there is a mismatch in the type of units available in the FSP as compared to the total requirement, with shortages existing for units from the transportation, quartermaster,

[2]TAA (Total Army Analysis) is the process used by the Army to develop its required force structure. TAA01 denotes the TAA for the Army program force for fiscal year 2001. See Sortor (1995) for a more detailed discussion of the TAA process.

ordnance, medical, composite services, aviation, and adjutant general branches. As shown in Figure 2.3, transportation, quartermaster, medical, and composite services in particular would all have significant shortfalls.

An MRC in northeast Asia (NEA) would also call for a faster deployment of support forces, and thus increase the need for active component support units. Even if M = C, in the NEA case the active component would need to deploy 67,000 active component support personnel in order to meet the desired force closure dates and readiness levels. For an MRC requiring reinforcement of combat and support forces (up to a total of eight combat divisions) or for two near-simultaneous MRCs, the Army would need to draw on support units from the less ready reinforcing and strategic reserve forces.[3] This could make additional reserve component personnel available (virtually all CONUS-based active component support units are in the FSP). These non-FSP reserve component units receive much less priority in peacetime, are at a lower readiness level, and consequently would require more resources and time for postmobilization training.

Both the ARNG and the USAR give their reinforcing support forces a lower priority for resources than their high-priority units. The ARNG, for example, plans to man its high-priority Enhanced Readiness Brigades and FSP units at 95 percent but to man the lower-priority units at 85 percent. These lower-priority units also receive less training assistance from active component units and full-time reserve personnel. Even high-priority reserve component units find

[3]Many have questioned the validity of the requirements that result from the TAA process. During the course of an earlier analysis we did some independent FASTALS model runs and compared the results to the deployment of forces for ODS. The results indicated that while FASTALS called for more Army forces than were deployed to ODS (primarily transportation, engineers, composite services, and medical), taking into consideration the specific features of ODS (extent of host nation infrastructure and support, no air threat, no large-scale construction of facilities) as compared to the planning scenario explained virtually all of the differences between the FASTALS requirements and the forces ultimately deployed to the theater. FASTALS also called for the forces to deploy more quickly (about 30 days faster) than indicated in the ODS TPFDD (Time Phased Force Deployment Data). These results would indicate that the force requirements from the TAA01 cases may call for forces to deploy somewhat more quickly than was the case for ODS and call for more forces in some branches than might be required under some sets of circumstances. See Sortor (1995) for a discussion of other statements of requirements and comparisons to FASTALS results.

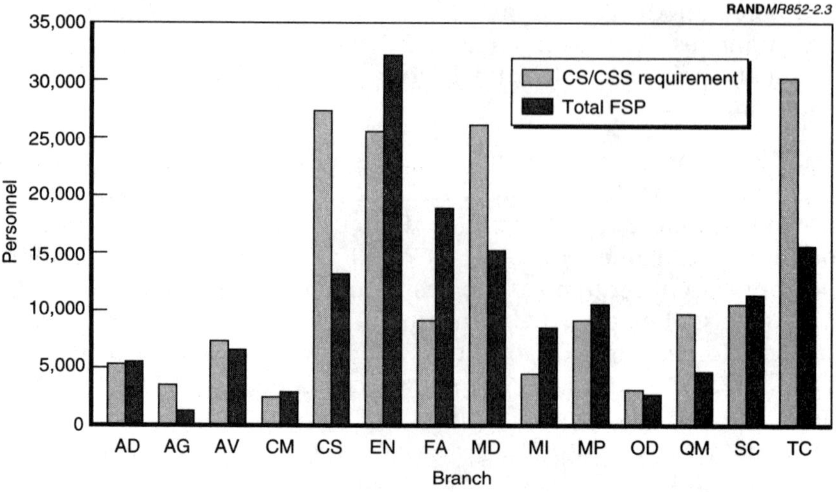

Figure 2.3—Comparison of Support Force Requirements and FSP Capabilities

it difficult to meet their readiness goals. Less than a third of the priority support units participating in the Army's Bold Shift program in 1993 were able to meet their peacetime training goals, and on average, the units reported they would require about 20 days of post-mobilization training before they would be prepared to deploy.[4] Readiness reports of the last couple of years indicate only slight improvement in readiness in the high-priority units.

Personnel readiness—having sufficient numbers of qualified people available to mobilize and deploy—is the primary constraint for these types of reserve component support units in meeting early-deployment requirements. Priority support units in 1992 and 1993 had only about 75 percent of their personnel skill qualified. In the Persian Gulf War, these units were brought to required readiness levels by cross-leveling personnel between units and assigning additional per-

---

[4]Bold Shift was a 1992 pilot program for improving the readiness of selected high-priority Army reserve component units. It included individual and unit training initiatives, new concepts for field training, and closer ties between active and reserve component units. See Sortor et al. (1994) for a more complete description of the Bold Shift program and its results.

sonnel from other sources. At that time, however, the Army enjoyed a much larger, more robust, well-trained, and well-resourced force to draw from. That is likely not to be the case in the future.

In summary, our analysis showed that under the assumption that only four to five divisions are needed for each major regional contingency, the ten-division combat force would be adequate even when judged against a scenario with two nearly simultaneous contingencies of this size. In contrast, however, it did not appear that the planned support force structure would provide the number of units at the needed readiness level to support the same simultaneous contingencies. The FSP would support, with some degree of risk, a single modest-sized contingency but not two nearly simultaneous contingencies. While support units other than those in the FSP exist in the planned forces, their lack of priority for resources would not permit them to be ready to deploy in time.

The Army has also completed a new total Army analysis that resulted in somewhat larger support requirements than those considered in our analysis. To satisfy part of the increased shortfall in CS and CSS for the two nearly simultaneous MRC case, the Army has proposed reconfiguring some of its existing support units and converting some ARNG combat divisions to CS or CSS units. These actions are expected to reduce the shortfall in support forces to an acceptable level. However, these actions would not come to fruition until well after the turn of the century. In addition, these actions do not address directly the potential shortfall in active support forces for the first MRC that would occur if mobilization were delayed or if significant numbers of active component support personnel are engaged in an OOTW and are not available immediately when the first contingency occurs.

Figure 2.4 shows the early-deploying active component requirements for selected branches under two mobilization assumptions and for the deployment timing from the more demanding SWA scenario discussed above. In Figure 2.4 we compare the requirement by branch to the active component contingency support unit capability in the FSP. Note that in this case even if M = C, there would be shortages in quartermaster and transportation capability in the theater during the initial deployment phase before reserve component units could be mobilized and deployed.

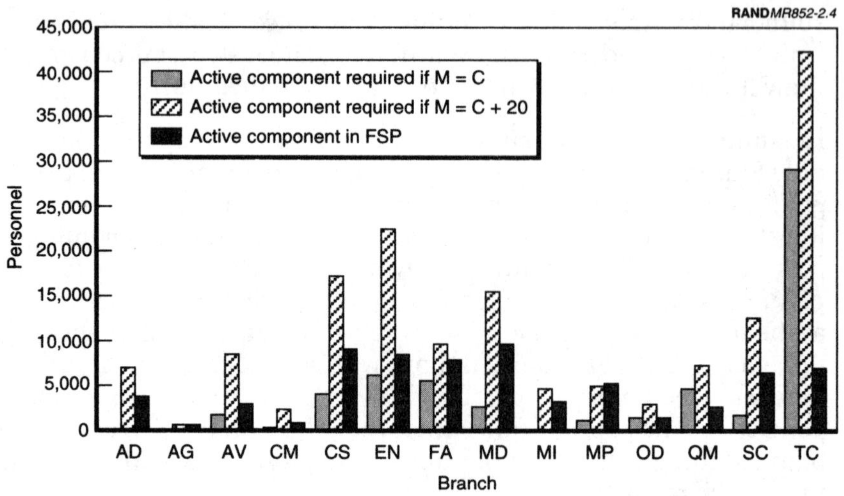

Figure 2.4—Active Component Contingency Requirements and FSP Capability

While personnel readiness is the primary constraining factor in reserve component readiness, active units experience a different problem. With a smaller force structure and growing demands for operations short of major regional contingencies (OOTW), the Army may not be able to maintain a contingency force that is sufficiently trained and ready to deploy to a major regional contingency that may occur with little notice. The BUR explicitly identified military operations other than war as objectives to be addressed by the armed forces. Recent examples include operations carried out in Somalia, Haiti, Macedonia, Bosnia, and for many years in the Sinai. The number, duration, and characteristics of such operations other than war could place demands on the Army force structure that would degrade the readiness of the active component forces below desired levels. Many of the units needed for typical OOTW missions are the same units required early in an MRC and, as will be discussed later, are units that exist in very limited numbers in the active component.

The research discussed in the remainder of this report examines a range of such operations and considers alternative approaches to them, including possible use of the reserve components, to develop ways to meet their requirements while preserving readiness for MRCs. First, however, we will look at the possible effects of forward presence on force structure and on the Army's ability to rapidly deploy to a major regional contingency.

FORWARD PRESENCE

Current planning envisions the Army keeping elements of at least three divisions overseas (two in Europe and one in Korea) for the foreseeable future. The 25th Infantry Division will remain in Hawaii. Each of the forward deployed divisions, however, will have only two of its brigades in the theater, with the third brigade of each in the CONUS. One infantry brigade will remain in Alaska. All of the other brigades, including the only two active component armored cavalry regiments, are stationed in the CONUS. The forward deployed forces also include about 50,000 personnel in support units positioned with the combat forces.

Since these forward presence requirements are fully considered in the force structuring process, we do not believe that they will place any added or unforeseen constraints on the Army's ability to deploy forces to MRCs (Sortor, 1995). In addition, the Army could, as in the case of the Persian Gulf War and some recent OOTW, determine that selected forward deployed units could be made available for an MRC or OOTW if required. Such a determination would be made based on the situation in the area of responsibility (AOR) of each of the forward deployed units and the criticality of the need elsewhere. For example, units may be available for an MRC outside of the AOR if there is sufficient stability in the AOR; however, availability of units to deploy outside their AOR could be severely hindered if there is political or economical instability or increased military hostilities in the AOR.

While "planned" forward presence is considered in force structure and contingency planning, "unplanned" forward presence is not. Military options may be limited by unplanned forward presence (such as keeping units in Panama longer than expected after Just Cause or deployments of Patriot units to Korea). Unplanned forward presence could impact MRC readiness in several ways. It is possible that while units are deployed on an unplanned forward presence mission, they would not be able to train to a required standard for an

MRC. Units that are forward deployed may experience greater levels of practice in their MOSs compared to their peers in the CONUS; however, they may have much less opportunity to train as part of a combined arms team. This lack of combined arms training would lead to a reduction in the Army's ability to maintain adequate forces ready to deploy on short notice to an MRC.

A unit's inability to train to standard while deployed on a forward presence mission would have only a short-term effect on readiness; that is, once a unit returns to CONUS, combined arms training is again possible. However, unplanned forward presence could have a more severe and long-term effect on force readiness. These missions can lead to repeated deployments and overseas tours of duty for soldiers. This increased OPTEMPO during peacetime has a negative impact on soldiers' morale—for example, increased time away from spouses, children, and family. If these missions have a long-term negative impact on soldiers' morale, the Army may not be viewed as a positive long-term commitment by soldiers, and these trained and qualified soldiers may not re-enlist.

Unplanned forward presence is simply one type of operation the Army performs that is short of a major regional conflict. Operations short of MRC and the problems associated with performing them may become even more prevalent in the future if the BUR proves correct in suggesting that new dangers will call for greater participation of Army units in operations short of a major regional contingency. Certainly, the experience over the last few years would support this thesis. In our analysis we define an unplanned forward presence, as well as other operations short of major regional conflicts, to be an operation other than war. More precisely, for purposes of this report, if an operation is not part of the current force sizing strategy—that is, two MRCs and planned forward presence—we will categorize it as OOTW.

OPERATIONS OTHER THAN WAR (OOTW)

Current strategy and guidance lays out a number of specific objectives for the armed services that could demand significant Army involvement in military operations other than war (MOOTW).[5] These objectives were presented in the *Report on the Bottom-Up Review* and included

- "Prepare U.S. forces to participate effectively in multilateral peace enforcement and unilateral intervention operations."
- "Use military-to-military contacts to help foster democratic values in other countries."
- "Protect fledgling democracies from subversion and external threats."

Currently the Army is involved in a number of operations like these. It is participating, or has recently participated, in operations in a number of locations (Rwanda, Somalia, Haiti, Macedonia, and Bosnia are examples) and has personnel (both active and reserve component) in numerous locations assisting in the training of forces of other countries and in military contact programs. The Army has since 1982 maintained an infantry battalion and a support battalion in the Sinai as part of the Multi-National Force and Observers mission (MFO-Sinai). The reserve components regularly deploy on overseas training missions that involve construction activities and medical support for people in foreign countries.

What kinds of operations are included under operations other than war? Army FM 100-5, *Operations,* 14 June 1993, defines OOTW as military activities during peacetime and conflict that do not necessarily involve armed clashes between two organized forces. FM-100-23, *Peace Operations,* includes support to democracy (peacemaking, peace building, and preventive diplomacy), peacekeeping, and peace enforcement. FM 100-19, *Domestic Support Operations,* describes another set of activities to include disaster assistance, environmental assistance, law enforcement, and community assistance. FM 100-16, *Army Operational Support,* describes MOOTW as military OOTW in which the Army conducts operations as part of a joint force. These operations may include support to U.S., state, and local governments, nation assistance, disaster relief, security and advisory assis-

[5]See the *Report on the Bottom-Up Review* for a complete discussion of the objectives (DoD, 1993).

tance, technical assistance, counterdrug operations, antiterrorism, support to domestic civil authority, and peacekeeping operations.

How do these operations differ from traditional combat operations as might be expected in an MRC or a general war, and do they place different demands on Army forces? One study (see Taw and Peters, 1995) concluded that these operations did present a unique set of challenges and identified at least two factors that were different in OOTW as compared to more traditional military missions. First, the study concluded that "most OOTW may have political goals that are much more complex and subtle and that infuse military decision-making at the most picayune levels of detail." Second, they concluded that OOTW, unlike more traditional missions, "do not move linearly from one set of tasks and objectives to another in a predictable fashion" but instead may move rapidly from peacekeeping to coercive measures and back to cooperative actions. There may be a lack of clarity of political objectives and the operational environment may be static, as in an operation like the MFO-Sinai, or dynamic, as in the case of Somalia and Bosnia. These differences place unique demands on force structure, equipment, doctrine, and training. In the next chapter we will consider these and other aspects of OOTW with an eye to understanding the military forces required for these operations and whether the planned Army force is capable of answering the demands of OOTW and an MRC.

For purposes of this analysis we will include under the rubric OOTW any operation involving military forces other than engagement in combat operations in a major regional contingency or general war. As will be discussed later, they may be in the United States or overseas, involve limited combat operations or not, and be under the auspices of the UN or coalition or be unilateral. Most of these operations do not place a severe constraint on Army capability to deploy to an MRC, as they tend to be short, small, and in need of general as opposed to very highly specialized capabilities. The Army has the flexibility to choose from a wide range of types of units or personnel to meet many of these limited demands. There are cases, however, in which the operation either is large and lengthy or demands a large proportion of a very specialized and limited capability. These cases place the greatest stress on Army force structure. In the next chapter we examine a number of operations conducted by the Army since 1975 and the characteristics of the forces used.

Chapter Three

HISTORICAL ROLE OF ARMY FORCES IN OOTW

OOTW are not new to the Army. The Army has been conducting these types of operations over its entire history. The Army is the DoD executive agent for domestic federal disaster response.[1] It also has an interest in domestic response by states even in cases where a federal response is not required because the National Guard is the first level of response to state and local emergencies. Thus even a state response may affect the availability and readiness of the ARNG units for federal missions, either OOTW or an MRC. Further, the Army has capabilities that make it uniquely suited to respond to OOTW requirements both domestic and overseas.

The Army is responsible for the land defense of the CONUS and thus has in place the C^3 and logistics structure needed to respond quickly to many domestic emergencies, whether disaster relief, riots, insurrection, or counterdrug operation support. It has many other capabilities that make it the desired response to overseas as well as CONUS OOTW needs. It has the premier capability for ground and helicopter transport, engineering and construction, water purification and distribution, medical care in austere environments, and large, demanding logistics operations. For these and other reasons, the Army can expect to continue to get the call to respond to most OOTW in which the United States finds itself. The question is, does the planned Army force structure provide the capabilities needed to respond to OOTW and an MRC? We will look to history and to cur-

[1]See Schrader (1993) for a discussion of the Army's role in domestic disaster response.

rent policy for an indication of the types of OOTW the Army may be called upon to accomplish and for the forces they may require.

Between January 1975 and June 1990 the Army participated in 49 operations, each involving 50 or more CONUS-based U.S. Army soldiers (USACAA, 1991).[2] Twenty-two of the operations were overseas and involved over 7,252,794 man-days (77.4 percent of the total) and 27 were conducted in the CONUS and involved 2,117,448 man-days (22.6 percent of the total).[3] Note that while these are large in absolute numbers, they represent only a very small fraction of the Army capability over this 15-year period—less than a quarter of a percent of the total active man-years. As will be discussed in greater detail later in this report, this indicates that, while important, OOTW requirements in the past have not been a major consumer, nor have they, for the most part, stressed Army capabilities.

Table 3.1 shows the proportion of the total man-days identified by type of operation.

Prior to Operation Joint Endeavor in Bosnia, Just Cause, a combat operation, was the largest of the OOTW-type operations in terms of the number of soldiers involved at any one time, with 10,566 soldiers deployed for a total of 443,772 man-days. But the largest operation, in terms of the total number of soldier man-days involved over the period, was the Multi-National Force and Observers (MFO) in the Sinai, a peacekeeping operation; from March 1982 to March 1990, it involved 15,991 soldiers for a total of 2,812,589 man-days.[4] The second largest was a nation-building operation (Joint Task Force Bravo) that, from August 1984 to June 1990, involved 11,168 soldiers for 1,323,323 man-days.

---

[2]The Force Employment Study (FES) was conducted by Concept Analysis Agency (CAA) to provide a historical data base by analyzing instances of Army forces employment during the post-Vietnam pre-ODS period. It identified 49 operations conducted between January 1975 and June 1990, each involving 50 or more CONUS-based soldiers, in which the Army participated. These 49 operations were grouped into nine types of operation.

[3]See the appendix for a more detailed description of these operations and the force structure templates that were derived from these data.

[4]The Army has had forces, approximately two battalions (one infantry and one support), committed to the MFO Sinai since 1982. This commitment is expected to continue into the foreseeable future.

Table 3.1

Employment of U.S. Army Forces (1975 to 1990)

Operation Type and Number	Percent of Total Man-Days
OCONUS operations	
Combat: 3	5.5
Peacekeeping: 1	30.0
Show of force: 1	.5
Security augmentation: 6	11.0
Nation building: 7	21.9
Humanitarian assistance: 4	8.5
OCONUS total: 22	77.4
CONUS	
Disaster assistance: 18	4.8
Support to law enforcement: 4	2.7
Refugee resettlement: 5	15.1
CONUS total: 27	22.6

SOURCE: USACAA (1991).

In evaluating the impact of OOTW on the total force, several variables need to be included, since the force structure implications of an operation depend upon a number of different factors. The number of personnel simultaneously engaged in an OOTW and the particular skills required affect a given force structure's capability to support an MRC. For example, an operation that requires five military police companies to be employed for one month will have a different effect on the availability and readiness of the overall force than will an operation that requires the same number of man-days of effort but with a single military police company deployed for five months or one that employs infantry, rather than military police, companies.

Force structure is affected by the overall length of the OOTW and two related factors—how long specific personnel are deployed and how frequently personnel rotate. Consider a case when only a single company is deployed at any one time. As the duration of the operation increases, the company could be in theater for a greater amount of time. Remaining in theater could affect the company's readiness for other missions (e.g., the unit may not be permitted to practice gunnery skills), and this could impact force readiness. As the length of time for the operation increases, so too does the need to rotate personnel. The rotation of personnel will involve more forces in training and transit than if no rotation was necessary.

Past deployment figures show that rotation length varied across operations. Table 3.2 gives examples of the number of troops deployed

Table 3.2

Force Required for Selected Operations

Operation	Forces Required and Typical Length of Unit Deployment	Total Man-Days
Combat operations		
TA battery—Lebanon	33 soldiers for 124 days	4,092
Urgent Fury	6,816 soldiers for 10 days	68,160
Just Cause	10,566 soldiers for 42 days	433,772
Peacekeeping operations		
MFO-Sinai	Inf Bn(+), 6-month rotations	2,812,589
Show of force		
Golden Pheasant	Inf Brigade for 14 days	44,366
Security augmentation		
MP ODT—Philippines	Various units on 14–21 day rotations	44,367
Nimrod Dancer	Mech Inf Reg for 210+ days	343,434
Nation building		
JTF Bravo	Various units on 120-day rotations	1,323,323
Fuertes Caminos	About 400 soldiers, 17-day rotations	181,686
Humanitarian assistance		
Incident in Guyana	270 soldiers for 15 days	4,050
TF Crosby	333 soldiers for 31 days	10,323
Eniwetok cleanup	330 soldiers on 120-day rotations	373,107
New Life	2,135 soldiers for 193 days	412,055
Disaster assistance		
Tornado cleanup, KY	69 soldiers for 4 days	276
Snow removal, OH	510 soldiers for 9 days	4,590
89 CA earthquake	2,805 soldiers for 29 days	81,345
Hurricane Hugo, SC	2,024 soldiers for 38 days	76,912
Law enforcement support		
ATC strike	Up to 225 soldiers for over 1 year	147,501
Hawkeye, Virgin Isles	1,429 soldiers for 48 days	68,592
Refugee resettlement		
Indochinese (2)	About 3,500 soldiers for 200+ days	849,036

SOURCE: USACAA (1991).

(at the peak of the operational deployment) on various operations, the length of time the units were typically deployed, and the total man-days involved for operations conducted between January 1975 and June 1990. Some operations may have deployed one unit or group of units for the entire operation, while in other cases different units were rotated into the operation over an extended period.

The flexibility in meeting the force requirements for different types of OOTW should also be accounted for in force structure implications. Specifically, we need to ask how time critical the various parts of the operation are to the whole operation, as well as how the length of the planning horizon for the operation could affect the force. Some operations require a certain number of man-days of effort which can be applied either more or less rapidly. In road building, for example, the various parts of the road can be built one after the other or a number of sections can be built at the same time, with only a small change in the total number of man-days of effort required but a large difference in the size of the force deployed at any one time. In addition, if the building of one section of road is delayed for some reason, it may have only a minor impact on the building of other sections. In these cases the force requirement could be altered to take account of the changing availability of units and the other requirements placed upon them without having a significant impact on the success of the operation. These factors are particularly relevant in determining when reserve forces may be employed rather than active forces.

The ability to draw on personnel and units from both the active and reserve components for OOTW has clear implications for the Army force structure. For almost all operations undertaken by reserve component personnel it would be feasible to use active component personnel instead; there are some operations that predominantly employed active personnel but could have used reserve forces. The only operations in which there might be problems using active component personnel are those involving the National Guard under state control.[5]

---

[5]This discussion ignores, of course, force structure limitations that would interfere with use of active versus reserve components. For example, most of the Army civil affairs and psychological operations capability is in the reserve components and thus currently limits the extent to which active component units can take on these missions.

However, there are several types of operations in which reserve component personnel probably could not be used to replace active component personnel. Consider, for example, a combat operation like Just Cause. The high level of risk, the need for operational security, and the urgency of such an operation will influence the decision on whether to use reserve or active personnel. Therefore, the past tendency to use active or reserve personnel to undertake particular types of operations needs to be considered.[6]

In summary, the following force requirement characteristics are key to determining how an OOTW is likely to affect the Army's readiness and its ability to quickly respond to an MRC:

- the number of personnel simultaneously engaged in the operation,
- the skills of the personnel (or force composition),
- the length of time a unit or individual is deployed,
- the total length of the operation and the need to rotate the personnel or units,
- how time critical the various parts of the operation are to the complete operation and the length of the planning horizon,
- the risk of combat, and
- the past tendency to use active or reserve component personnel to undertake that type of operation.

Based upon the recent historical evidence, not only are OOTW not new missions, some of them can call for a significant commitment of Army resources over an extended period. They are also now receiving even greater emphasis. The Army recognized in a recent publication describing the conceptual foundations for the conduct of future operations into the 21st century that strategic interests have increased both the number and the range of OOTW the Army will be required to perform in the future, and that OOTW and low-intensity conflicts will comprise most of the conflicts involving the Army

---

[6]This does not mean that our analysis will be limited to assuming this tendency continues, but that our initial look at the force-structure implications of OOTW will assume it continues.

(TRADOC, 1994). The BUR envisioned the possibility of the Army being involved in larger interventions, as well as performing smaller OOTW missions as it has done historically. The BUR established that such a larger intervention could involve up to two Army divisions, Marine and Air Force units, and support forces totaling 50,000 combat and support personnel (DoD, 1993).

Figure 3.1 shows the proportion of forces (measured in terms of man-days of effort) by branch that were committed to operations between April 1975 and June 1990 based on data in the FES.

Figure 3.2 further aggregates this data to show commitment (in terms of the percent of total man-days of effort) summarized over combat, combat support, and combat service support.

The OOTW considered vary in size from less than a single company to two divisions. The force compositions also vary between operations centered around combat units, those centered around combat support units, and those centered around combat service support units. Figure 3.3 summarizes the force requirements with operations grouped as in the above discussion. The forces used in the past

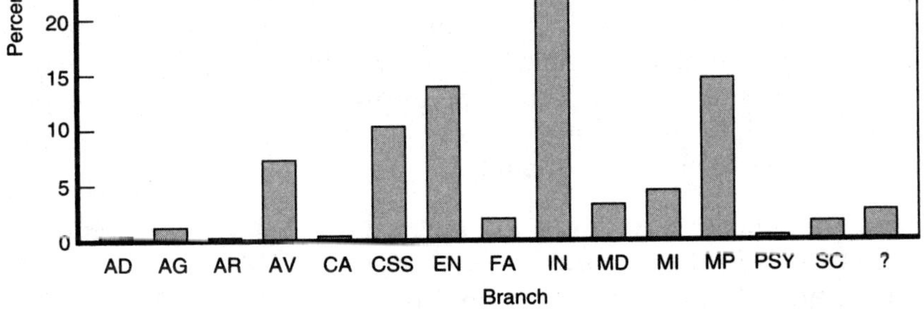

Figure 3.1—Historical Commitment of Forces to OOTW (Measured in man-days of effort)

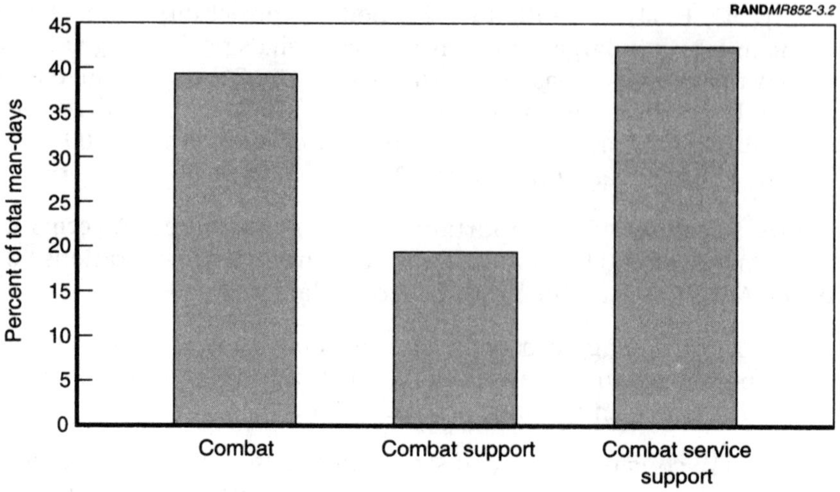

Figure 3.2—Type and Mix of Forces Required for OOTW

combat, armed intervention, and peace enforcement operations considered above were built around infantry combat units. However, in the future such operations could be built around a different mix of combat units. Similarly, peacekeeping could be undertaken by MPs rather than infantry and security augmentation by infantry rather than MPs, as indicated in Figure 3.3.

The deployment of an infantry or mechanized combat force of about a division could be undertaken as a combat operation, a peace enforcement operation, or an armed intervention operation. Thus these three types of OOTW operation can be represented, in terms of force requirement, by a single case and would also represent the high end of a possible show of force operation. It should also prove possible to extend this case reasonably simply to cover the deployment of a two-division intervention force.

Similarly, the deployment of an infantry or MP force of between battalion and brigade size could be undertaken as a show of force, security augmentation, peacekeeping, or law enforcement operation. The deployment of a battalion-sized or possibly brigade-sized force centered around CSS, EN, MP, or IN units could be undertaken as a

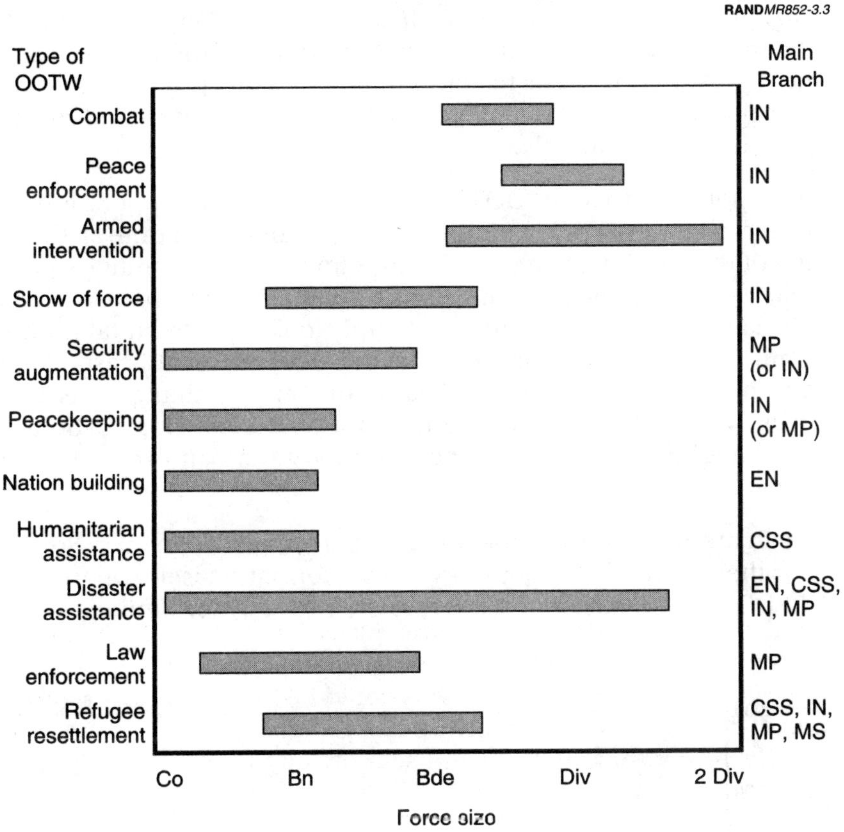

Figure 3.3—Size and Composition of Different Types of OOTW

nation building, humanitarian assistance, disaster assistance, or refugee resettlement operation.

These three cases, and their force requirements, can be considered representative of the wide range of OOTW and of the capabilities the Army may be expected to provide for future operations. The force requirements have been put together so that if they can be met, then the Army should be able to meet the requirements of most types of OOTW. They can be scaled up or down as required to cover the variability in the size of force actually required for a given type of OOTW. These three cases do not include operations that require relatively small numbers of specialists, such as domestic emergencies, the deployment of a target acquisition battery to Lebanon, or the deployment of Patriot batteries to South Korea. These represent a different level of capability (more individual as opposed to unit) and need to be considered separately.

An important aspect of OOTW is the duration of the operation and the fact that the force requirements may change over time as the nature of the mission changes. The announcement that military police units would soon be replacing units from the 1st Armored Division in Bosnia is a very recent example. Simply looking at the total number committed to the operation in a static view misses this aspect and fails to indicate how the force requirements may change over time. To illustrate, Figure 3.4 shows for Operation Restore Hope in Somalia the changing nature of the operation over time and the forces required.

The figure shows the number of Army personnel deployed in Somalia over the course of the operation. This information is drawn from the TPFDD (Time Phased Force Deployment Data) for Somalia, which

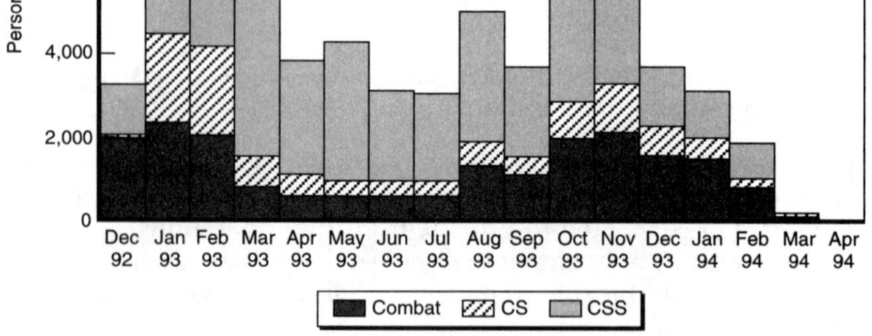

Figure 3.4—Personnel Deployed in Somalia

has been used to determine the number of personnel and units needed in the OOTW and the rotation of these units.[7] The initial buildup of the deployed force is followed by a scaling back in the commitment. Then, following Warrant Officer Durant's capture, the force again expands as the heavier mechanized units are deployed.

OOTW cover a wide range of types of operations, and each may require a different mix of forces. In many cases, however, there may be a great deal of flexibility in how the force is configured and the exact mix of units or personnel that might be called upon. Two previous studies have identified the need for a force structure that is flexible and easily tailored to meet the demands of particular OOTW (Taw and Peters, 1995, and Davis et al., 1997).[8] In addition, as we will see later in this analysis, the mix of skills needed will often change over time as the operation unfolds. Also, Taw and Peters found that OOTW can change rapidly from one type of task to another rather than follow a predictable linear progression of tasks as seen in more traditional military operations. Further, the Army forces are organized and trained for employment in major combat operations. As identified in Davis et al. (1997) for medical units, and as we will see in other cases, OOTW most often require forces to be tailored for the specific operation in ways that are nondoctrinal and that were not envisioned when the units were configured or trained.[9] These aspects place a greater burden on the unit leaders and commanders than might be the case for an MRC.

The next chapter presents a methodology for exploring the requirements for OOTW and their possible effects on the readiness of Army forces to quickly deploy to an MRC while simultaneously engaged in one or more OOTW.

---

[7]TPFDD data obtained from FORSCOM were current as of November 3, 1994.

[8]See Davis et al. (1997) for the results of a study focusing on the role of the Army medical capabilities in OOTW and an examination of its role in Bosnia, Somalia, and Haiti.

[9]In the past, Army doctrine, force design, unit equipage, and training have been centered almost exclusively on warfighting missions.

Chapter Four

MODELING THE EFFECTS OF OOTW

Our research is directed at the question, How might the conduct of operations such as peace enforcement, humanitarian assistance, peacekeeping, and lesser regional contingencies influence the readiness and availability of Army forces to deploy to an MRC? This question is addressed in two parts. The first part looks at how OOTW deployment affects the availability of individual units and focuses on how long units take to prepare for deployment, deploy, engage in OOTW operations, redeploy to home station, and complete recovery and training so the unit is ready for possible deployment to an MRC. The second part examines the effect of OOTW on the force structure as a whole and is focused on the ability of a given force structure to meet the OOTW requirements and to simultaneously maintain a sufficient number of ready and available units to deploy, should it be necessary, to a single MRC. It considers, in addition to the factors involved in unit availability, the need to sometimes deploy multiple units of the same type on lengthy OOTW operations and then use unit rotations to relieve units in the theater before the OOTW is completed. We will first illustrate the methodology for assessing these effects and then in the next chapter apply the methodology to deployments for Operation Restore Hope in Somalia.

UNIT AVAILABILITY

Engagement in OOTW has a number of effects on a unit's availability. For our purposes the effects are captured by assessing the period of time the unit is unavailable for deployment to a second operation. Units become unavailable for immediate deployment to an MRC at some point in their preparation for an OOTW deployment. They are unavailable while employed in the operation and remain unavailable while recovering from the OOTW deployment. Figure 4.1 displays these steps and depicts the status of a given unit over time as it moves from a ready status (i.e., ready and available for immediate deployment to an MRC) through involvement in an OOTW and then back to a ready status.

The time required to prepare for participation in an OOTW and the time required after return to the CONUS to regain ready status will vary depending on the type of unit, the length of the deployment, and the nature of the OOTW mission. For example, support units are more likely to carry out a task closely related to their task in an MRC; thus an OOTW deployment may have only a limited impact on their level of training. However, a long deployment could give rise to significant personnel turmoil when a unit returns to home station and have a negative impact on the level of collective training. It is difficult to identify the precise influence of these factors on the time required. However, approximate figures for the time required will be sufficient to indicate the magnitude of the potential problem.

The time taken to deploy the unit from its home duty station to the OOTW theater will depend on how far it has to travel and the mode of transport used. The time to deploy a unit by sea, from the CONUS to a foreign country, is likely to be significant. In some cases it may be possible for successive units rotated into an ongoing OOTW to use the equipment taken into theater by the first unit to deploy. This would allow later-deploying units to deploy faster, as they could

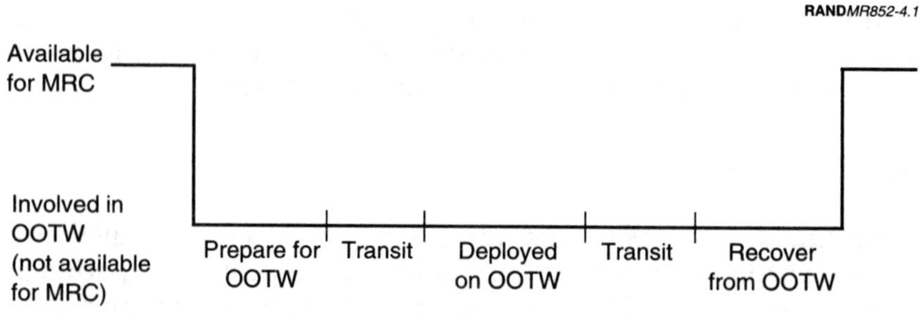

Figure 4.1—Unit Availability and Status

travel by air and join equipment already in the theater rather than wait for their equipment to arrive by sea. However, successive deployments using the same in-theater equipment might have implications for the amount of training needed and would raise issues associated with how the first unit to deploy would be reequipped.

When an operation requires units to be rotated, with one unit taking over the task of another, there will need to be a handover period. The length of time required will depend on the type of mission. An approximation of the time will be sufficient, particularly as it is likely to be short compared with the overall time to prepare, deploy, recover, and retrain the unit.

Figure 4.2 depicts the steps most units typically go through after returning from an extended OOTW deployment. These steps are not unlike those required to prepare for such a deployment.

Whether departing for or returning from a deployment, a unit must inventory and prepare its equipment. If deploying, the unit would prepare the equipment for shipment; if returning, it would unload and prepare the equipment for storage or use in training. In most cases the unit would schedule block leave so unit members would have an opportunity to take leave and spend time with their families. The unit would then go into a training cycle to regain combat-ready status. This might take only days or weeks for some units, months for others. Finally, the unit, if going into a ready status, would in-

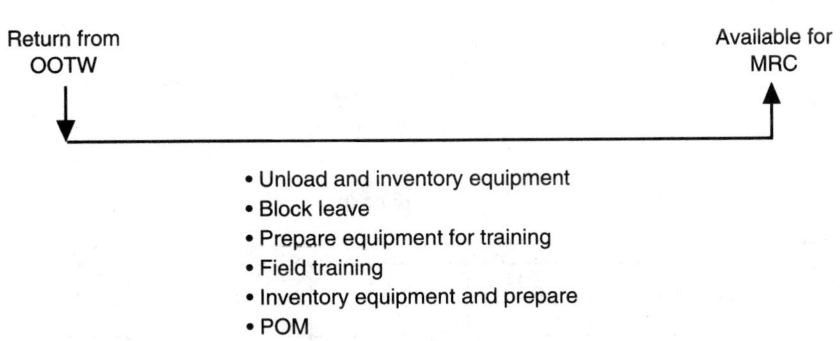

Figure 4.2—Steps Required to Recover Ready Status

spect and perform required maintenance on unit and individual equipment and take care of personnel actions (preparation of wills, shots, physicals, etc.) required in preparation for possible overseas movement (POM).

As discussed above, the time required to complete this cycle will vary widely depending upon the type of unit, the nature and duration of the operation, and the degree of similarity between the tasks performed during the operation and the tasks normally performed by the unit in an MRC. Figure 4.3 portrays a deployment and training timeline for an infantry battalion, patterned on the cycle typically followed by units deploying to MFO-Sinai.

In the case shown in Figure 4.3, the infantry battalion takes about five months for preparation and training for the OOTW mission. One month is required for the actual deployment and transfer of mission responsibility from the previous unit. Up to six months before deployment the unit begins some preparation and training in conjunction with its normal peacetime training activities. At about four months the unit will begin focusing almost exclusively on MFO training. The actual OOTW mission lasts six months, after which the unit returns to its home station. The unit schedules a month for block leave and routine processing before beginning a three-month intensive training cycle to regain combat-ready status. This illustrates the often-cited case that these operations, like MFO-Sinai for example, require about three units in the force in order to support one deployed.

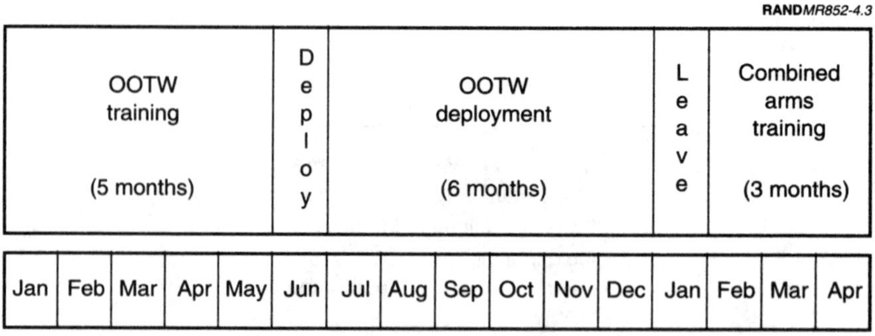

Figure 4.3—Infantry Battalion Deployment Training Cycle

The Center for Army Lessons Learned (CALL) conducted a study in 1995 to develop detailed timelines with tasks, showing best- and worst-case estimates for predeployment training for OOTW mission execution and postdeployment operations for combat, combat support, and combat service support units. The data for predeployment preparation and training ranged from 18 days to 10 weeks. For contingency-type operations, units would have only a short period of alert and little if any time for specific mission planning or training. The unit must rely on previously developed skills and the flexibility inherent in well-trained military units. For other operations, like MFO-Sinai, the operation is planned for in advance and the unit has a much longer period to prepare. In those cases it is likely to develop very mission-specific training events and take time to hone its skills in those particular activities. Figure 4.4 shows the process and timelines developed by CALL for the recovery phase.

Even in the limited number of cases CALL examined, there was fairly wide variation in time to recover. While different types of units (light infantry, armor, combat support, combat service support) have different requirements to complete to reach combat-ready status, and therefore take different amounts of time to recover from an OOTW,

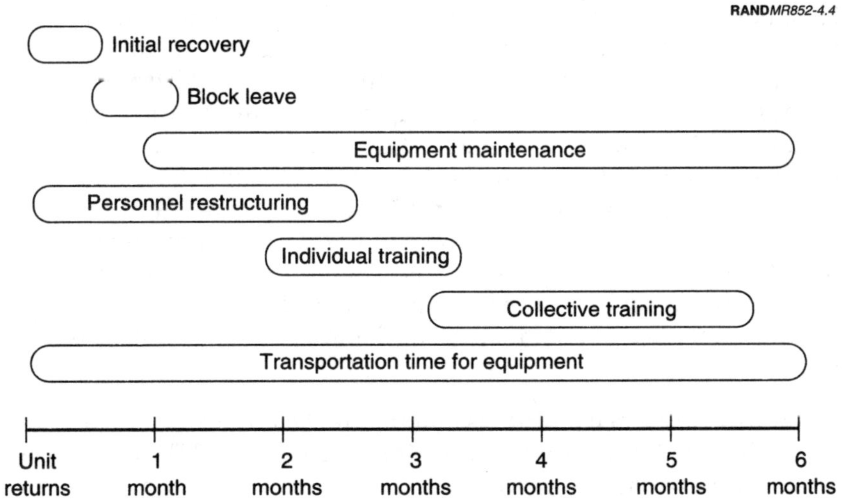

Figure 4.4—Readiness Recovery Process

CALL found a common set of tasks or phases that all units went through. These are not unlike those depicted in Figure 4.2. CALL found that many of these common phases required about the same amount of time regardless of the unit type. For example, unit recovery took one to two weeks, block leave was typically two weeks, personnel restructuring took about three months, and individual training took four to six weeks. Greater variation existed for collective training, depending on the type of unit. Combat units and some combat support units took eight to ten weeks, while other combat support and combat service support units took only two to four weeks of training. The two largest sources of variation were equipment related. Equipment maintenance took between one and six months after equipment reached home station. Transportation of equipment to home station also took from one to over six months. Some shipments took as long as eight to eighteen months, due to extreme delays or outright loss. In addition, a unit's equipment is sometimes left in the theater, so it must wait for replacements from new purchases or from depot stocks. This may take considerable time.

General Henry Shelton also identified these equipment-related issues among the readiness impacts of participation in OOTW (Shelton, 1995). His message cited examples like the 86th Combat Support Hospital: after leaving its equipment in Somalia, this unit had only recovered to 85 percent eighteen months later. In other cases, because of continuous operation in a harsh environment and because of the austere theater, equipment required extensive maintenance after its return to home station. Such factors can help cause particular units to take from as little as six months to as much as a year and a half or more to fully reach combat-ready status after return from an OOTW. These time estimates and experiences all are based on a peacetime business-as-usual situation. Clearly, given priority and a sense of urgency, units could regain at least partial readiness status and be prepared for deployment to a combat theater much more quickly.

CALL also developed a process and timeline estimates for an accelerated schedule for units being withdrawn from an OOTW for the purpose of redeployment to an MRC. This is not unlike the experience of a number of units that were engaged in Operation Restore Democracy in Haiti when they were notified of deployment to Kuwait for

Vigilant Warrior. Some units, like elements of the 7th Transportation Group, redeployed from Haiti to the United States, where they prepared for deployment and then quickly deployed to Kuwait. Other units deployed directly from Haiti to Kuwait. This process and timelines constructed from the CALL data are illustrated in Figure 4.5.

Figure 4.5 illustrates a situation where a unit is engaged in a OOTW but while deployed is identified for employment in an MRC and given priority for regaining deployability status. Note that in this case no time is given for block leave, personnel restructuring does not occur, and, with priority, time is greatly reduced for equipment redeployment and preparation. Equipment may not be brought to full mission-capable status, while only the essential tasks are performed to bring dead-lined equipment to operation. The reduced time to prepare the units may also reflect a training schedule that would only result in very limited training events (no live fire exercises for combat arms units, for example) and would rely on skills retained from past experiences. For combat arms units, these variations would result in minimum training times of 30 days, but 75 to 90 days would be needed to bring the unit to full combat-ready status. Combat service support units would take a minimum of two weeks but need up to two months in many cases to reach full combat-ready

RANDMR852-4.5

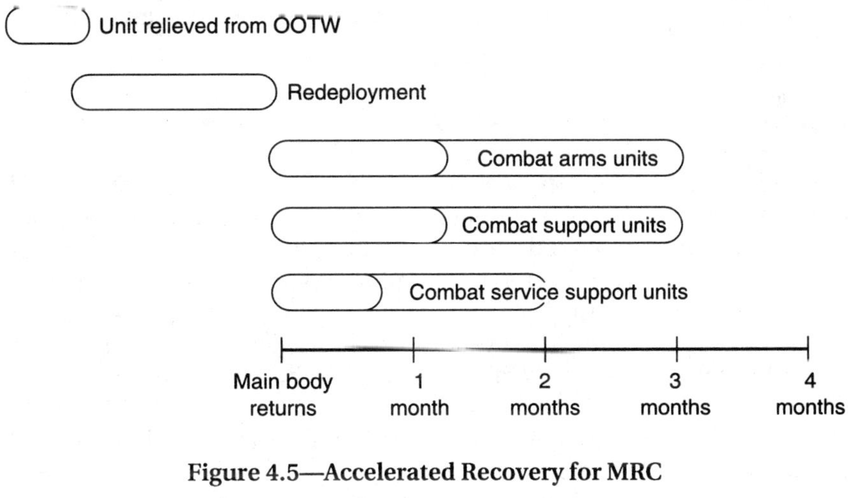

Figure 4.5—Accelerated Recovery for MRC

capability. Equipment availability and condition would be pacing items for many equipment-intensive units, particularly for the shorter timelines.

Note that this accelerated timeline assumes that the unit is identified for onward movement to an MRC at the time it is withdrawn from the OOTW. Once a unit commences the steps shown in Figure 4.4 for a normal peacetime recovery timeline, it is virtually impossible to reverse course and begin executing the accelerated schedule. Personnel who have departed the unit would be difficult to recall, and new personnel would extend the time needed for training prior to deployment. Equipment torn down for repair or overhaul and awaiting parts would not be immediately available either to support training or for shipment.

The above is an approach for defining the training and deployment cycle and timelines for individual units and for determining unit availability. With these estimates and variations to account for differences in the length of deployment and type of OOTW, we can explore the likely effect of OOTW on the availability of units to return to ready status in peacetime or, if an accelerated need is identified, to be ready to deploy to an MRC. The next question is force availability. Force availability is a function of not only unit availability in the context of the above discussion, but also the number of such units in the force, the number of units simultaneously engaged in OOTW, and the duration and rotation cycle policy for units deploying to an OOTW.

FORCE AVAILABILITY

We will use a simple model to explore force availability issues and options. The simplest case relates the deployment time to the fraction of the force deployed. For example, if one-third of the force is continuously deployed on a rotating basis, then each unit, on average, will spend one-third of the time deployed. As the percentage of the force that is deployed increases, the units will spend a larger and larger fraction of the time deployed. This is the situation the Army is facing. As the size of the force is reduced, there are fewer units to share in OOTW deployments. At the same time, the number and duration of the OOTW missions are seen to be increasing. The result is that units spend less time at home station and longer and more frequent periods deployed. The force availability model is depicted in Figure 4.6.

Figure 4.6 depicts the relationship between the time units are at home station (H) to the fraction of the force deployed at any given time (F) and to the length of the deployment (T), given a long-running OOTW or ongoing OOTW demand. Thus, as the Army has gotten smaller, the fraction of the force deployed has increased and, as a consequence, the units spend less time at home. This situation is illustrated in Figure 4.7.

Note that in Figure 4.7, as the fraction of the force deployed increases from point F1 to F2, the time at home station, assuming a deployment time of T, decreases from H1 to H2. For example, assume there are six units in the force and the policy is that units should spend no more than three months deployed to an OOTW. If an OOTW requires only two units (one-third of the force), then each unit will spend three months deployed and six months at home before repeating the cycle. If four units are required, however, then each unit will still spend three months deployed but only one and a half months at home before repeating the cycle.

While lengthening the time deployed will lengthen the time at home, it does not affect the overall percentage of time the units and per-

RANDMR852-4.6

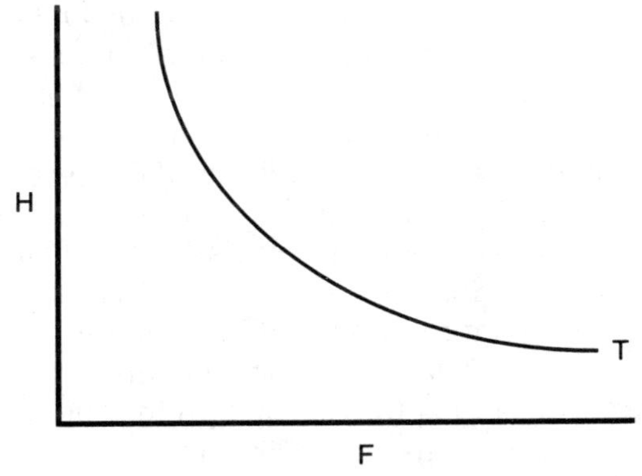

Figure 4.6—Force Availability Model

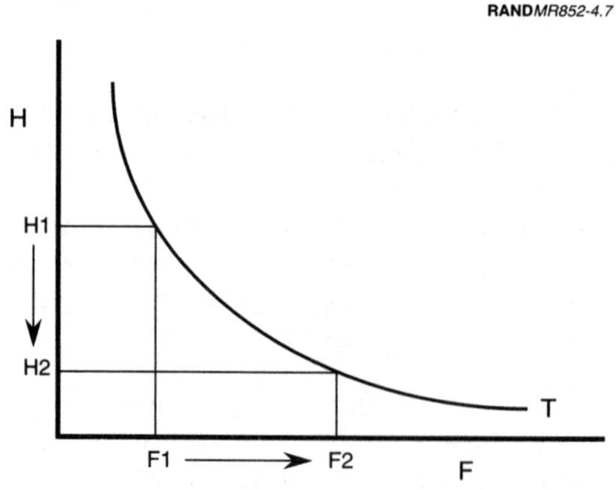

Figure 4.7—Increasing the Fraction of the Force Deployed

sonnel spend deployed if we assume units are rotated and all units participate in the deployments. Mathematically, the proportion of the time a force is deployed is equal to the proportion of the force deployed. The length of a unit's continuous deployment on an operation (tour length) is important, however, as it determines the number of months the unit is at home and thus able to retrain for its MRC mission, be prepared and ready to deploy to an MRC, or train for a subsequent OOTW deployment.

While desired tour length is one of many important factors in force design, when the Army is faced with conducting an OOTW with the force structure at hand, the *only* aspect of force availability and readiness that the Army can control is the length of time the units are to be deployed on a given rotation. The Army could leave the initial units in the theater of operations until the OOTW is completed. For short missions, as is often the case for disaster relief, the latter may be an acceptable solution. For longer missions in austere or adverse environments, such as Somalia, this is less acceptable. Certainly in cases like the MFO-Sinai rotation that has gone on for years, this is not an option. Figure 4.8 depicts graphically the conceptual tradeoff available in terms of varying the tour length in order to lengthen the time at home station for units and their personnel.

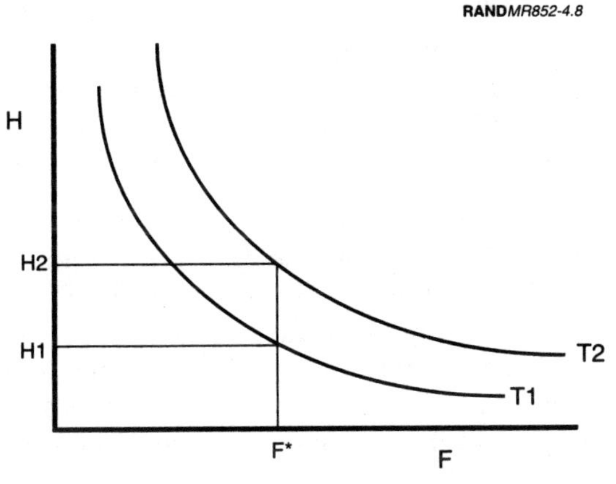

Figure 4.8—Deployment Tour Length Versus Force Availability

In Figure 4.8, if the rotation cycle for the fraction of the force F^* is increased from T1 to T2, then the time at home station is increased from H1 to H2. Take our earlier example, where there are six units in the force. Table 4.1 shows how long units would be at home, on average, depending upon the fraction of the force required for the OOTW and the length of the deployment tour.

In the case shown in Table 4.1, the length of time a unit can expect to spend at home station can vary from 30 months to as little as 1.5 months, depending upon the fraction of the force required for the

Table 4.1

Length of Time at Home Station (in months)

	Time at Home (Months)	
Fraction of Force Deployed	Three-Month Deployment	Six-Month Deployment
1/6	15	30
1/3	6	12
1/2	3	6
2/3	1.5	3

OOTW and the length of each deployment. This tradeoff becomes more of an issue as the time it takes units to train for an OOTW, deploy, redeploy, and retrain for an MRC becomes longer, consumes more of the time at home, and thus reduces the number of units at home that are "ready." This is illustrated in the following example.

In Figure 4.9 we show the case where there are nine units in the force and two are deployed at any one time on a long-running OOTW. The units are deployed for either three or six months. The time taken to prepare, deploy, redeploy, and retrain units is taken to be zero, two, or four months. The number of units shown available for an MRC is the minimum number available; because of rounding, at certain moments the actual number available may be higher by one unit, but it will never be lower.

The tour length is not important if the units are instantaneously available for any mission when they are at home station, assuming that no time is taken moving to and from the OOTW and to transfer mission responsibilities. This is unlikely, however, and most units will require some time to prepare for an OOTW, to move to and from the OOTW, to transfer responsibilities, and to recover and regain

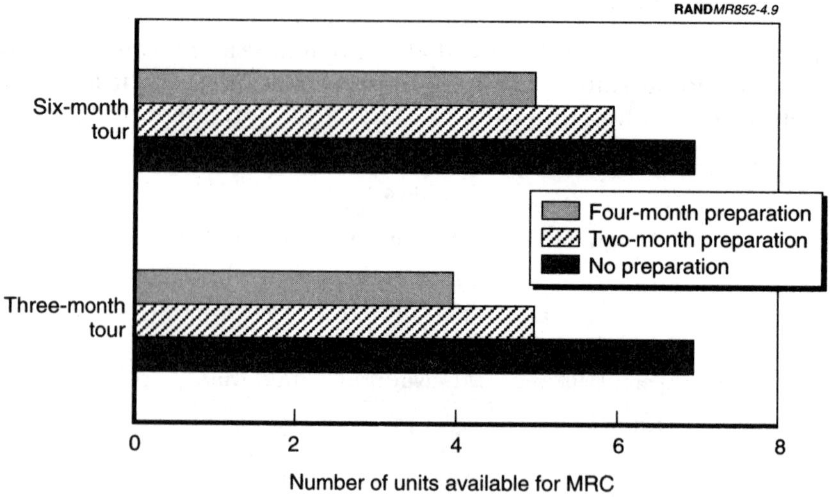

Figure 4.9—Preparation Time and Tour Length Effects

ready status for possible deployment to an MRC. As the time required to undertake these actions increases, the shorter deployment rotations result in fewer units at the ready, even though the overall proportion of the time any one unit spends deployed remains the same.

In the case where we are designing a future force, we can use this methodology to identify the number of units needed in the force structure to ensure that planned OOTW requirements can be met and that a sufficient number of units will always be ready to deploy to a major regional contingency should it be necessary. The force structure can be determined based on a desired tour length and rotation schedule defined in terms of how long units are to be deployed to an OOTW and on the desired time between deployments.

For example, take the case where the MRC requirement is for four active component units and three units are needed for a long-running OOTW. Assume the preferred rotation schedule is for a six-month rotation to the OOTW. Further assume the train-up cycle is similar to that shown earlier for the infantry battalion, e.g., a six-month train-up and transition for an OOTW and four months to return to ready status after an OOTW deployment. We then need four units at the ready for an MRC and three in the OOTW, with five units in some portion of the train-up cycle. Thus the force structure would need a total of 12 of these units in the active component to satisfy all of the stated requirements. This is illustrated in Figure 4.10.

The same methodology can be applied to cases where the limiting resource is individual specialists instead of units. The methodology could also be expanded to cope with cases where the limiting factor is equipment instead of units or personnel.

As noted briefly above, frequent participation in certain types of OOTW may have an adverse impact on units, equipment, or individuals over the longer term. For example, repeated and frequent deployments, or even occasional deployments to unpopular operations, may be expected to affect morale and reduce personnel retention. Therefore, there may exist a certain desired minimum time between deployments for a unit or individual. Such a desired minimum time between deployments can easily be represented in the analysis. In addition, the analysis can consider the percentage of

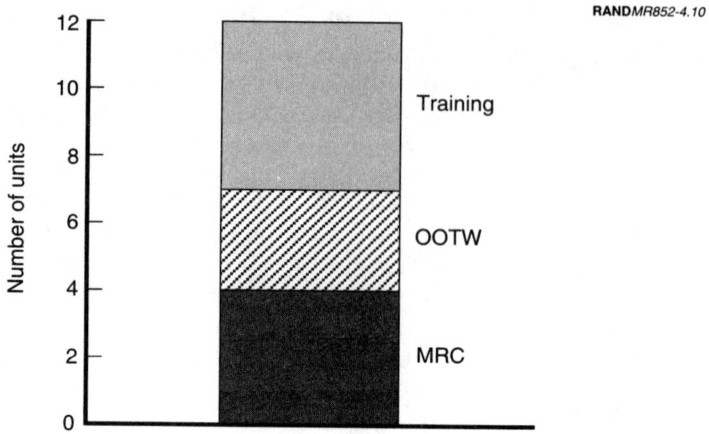

Figure 4.10—Force Structure Required for OOTW and MRC

time that individual units are likely to be deployed or involved in OOTW.

There may be other limits on which units can be dispatched to OOTW and their effects on force availability and readiness. For example, if all the field kitchens at a base were sent on an operation, then other units at that base might have their ability to undertake field operations and training severely limited or completely eliminated. Thus, different policy options may have an impact on the readiness of the units beyond those sent on OOTW deployments.

The analysis initially concentrates on cases where it is assumed that active component units are deployed on OOTW missions. Subsequently, we consider options for using reserve component forces. In the next chapter we use the above methodology to assess the implied readiness effects and their implications for the Army's MRC capability, using data from a recent OOTW deployment.

Chapter Five

EVALUATING THE EFFECTS OF OOTW ON ARMY FORCES

Using the estimates for preparation and recovery time from the CALL report discussed in the previous chapter, together with the actual deployment of units to Operation Restore Hope in Somalia, our previous work on MRCs (Sortor, 1995), and our model of unit availability, we shall now illustrate the impact of a single OOTW (Somalia) on the availability of the force. We shall also discuss how varying the length of deployment, time to prepare, transit time, and recovery time changes the overall force availability. First we show the impact of Somalia on the capability of the BUR force to respond to a single four- to five-division MRC.

Figure 5.1 shows the number of Army personnel deployed in Somalia over the course of Operation Restore Hope. This information is drawn from the TPFDD for Somalia, which has been used to determine the number of personnel and units needed in the OOTW and the rotation of these units. The initial buildup of the deployed force is followed by a scaling back in the commitment. Then, following CWO2 Durant's capture, the force again expands with deployment of armor forces.

As described above, individual units are considered to be involved in an OOTW, and hence not available for early-deployment to an MRC, while preparing to deploy, in transit, and recovering after the deployment. Therefore it is necessary to take account of these times when considering the Somalia deployment's impact on the availability of the Army. We have assumed that preparation to deploy and transit to Somalia takes 30 days. This is based upon an estimate of about 20 days for transit by ship together with 10 days of preparation.

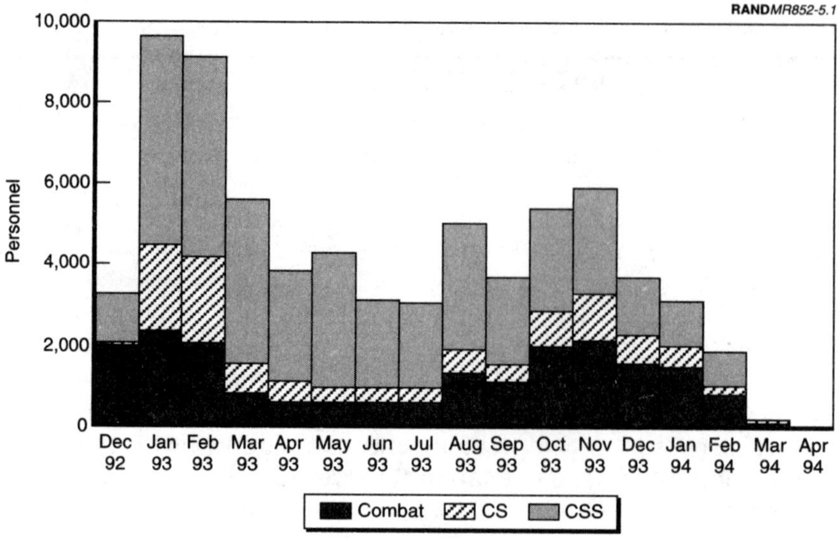

Figure 5.1—Personnel Deployed in Somalia

The 10-day preparation time is at the lower end of the range of preparation times for contingency operations (18 hours to 10 weeks) as identified in the CALL study. Thus the estimate of 30 days is very optimistic. Similarly, we have estimated the time taken for units to redeploy from Somalia and recover to full combat-ready status as

- Three months for light combat arms
- Four months for heavy combat arms and CS units
- Five months for CSS units.

These estimates are consistent with the timelines developed from the CALL data and reflect a peacetime business-as-usual priority for resources such as personnel replacements and access to training opportunities. The longer time for CSS units reflects their historically lower priority for resources and training activities as opposed to the need to support garrison activities and field training for other units rather than their own readiness recovery needs. Later we will explore options for reducing these time estimates and the potential effects on unit and force availability.

It should be noted here, however, that the recovery time is very dependent on the size of the unit and the nature of the particular operation. Even for a light combat arms unit, for example, the recovery time will vary greatly if we are talking about a platoon as compared to a battalion or a brigade and whether the unit had the opportunity to train while deployed. In the case of a platoon deployed to an OOTW, its time to recover in order to join the other elements of its brigade for deployment to an MRC would be expected to be very short. If the entire brigade was returning from an OOTW, the recovery time before it would be ready to deploy to an MRC would likely be closer to the middle to higher range of the CALL data. General Ronald Griffith, Army vice chief of staff, is reported to have said that the 1st Armored Division would need at least 90 to 100 days of training alone after withdrawal from Bosnia before it could be ready for deployment to a combat operation (*Army Times,* May 6, 1996).

EFFECTS OF PREPARATION AND RECOVERY TIME ON UNIT AVAILABILITY

Combining the above estimates for preparation, transit, and recovery time with the information on unit deployments from the Somalia TPFDD, we can calculate the number of personnel involved in the operation over time. This is shown in Figure 5.2.

Figure 5.2 shows the total number of Army personnel (combat, CS, and CSS) deployed in theater and, using our model of unit availability, the total number involved in the operation and therefore unavailable for immediate deployment to an MRC. As stated above, our assumptions for the time taken to prepare, transit, and recover are felt to be optimistic, so the figure understates the true impact of the Somalia engagement.

The peak number of personnel in theater occurs early in the deployment in January, yet the peak number of personnel involved in the operation occurs several months later in May and is 67 percent higher than the peak number in theater. Furthermore, in July, when the number of personnel deployed had dropped to just over 3,000, the number of personnel involved in the operation was over five times higher than the number in theater. This does not include units in the CONUS or elsewhere that may have supported the deployment to Somalia without themselves deploying.

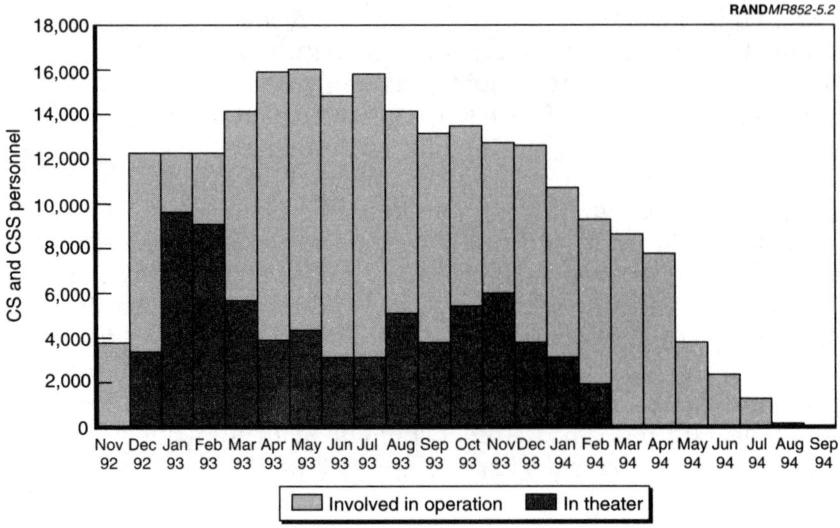

Figure 5.2—Total Personnel Involved in Somalia Over Time

If an MRC were to occur during July, for example, then it would be possible to deploy early in response to the MRC many of the units shown above as involved in the operation. This would certainly apply to most of the units that were preparing to deploy and to many that were recovering but to only a few, at best, of those in transit or deployed in theater. Many of these units would have a reduced combat readiness, as they would not have been planning on needing to deploy early to the unexpected MRC, and the resources necessary to accelerate their return to full combat readiness would not have been made available before the crisis occurred. Therefore, it is better to plan for a situation in which there are sufficient combat-ready units within the force to deploy early to an MRC without the need to deploy units which may well have a reduced combat readiness.

Several of the units that deployed to Somalia did not deploy their entire strength, as only part of the unit was required. However, the deployment to Somalia of part of the unit has an impact on the combat readiness of the entire unit and hence the unit's availability. The impact on the unit availability varied from unit to unit and depended on a wide range of factors, including exactly who deployed (which particular skills and grades), what equipment was deployed (the extent to which like-type equipment was available for training), the amount of support the nondeployed component was required to supply to the deployed component, etc.

Let us assume that if over 30 percent of a unit deployed to Somalia, then the unit as a whole should be counted as involved in the operation and unavailable for deployment to an MRC.[1] We choose 30 percent because it corresponds to the remainder of the unit being at less than 70 percent strength, which is C-3 or worse. Note that this assumes the unit was at 100 percent of strength prior to deployment. Many of the support units are organized at less than ALO 1 (full wartime requirement) and are undermanned even in comparison to their authorized strength. As a result, many units would be at much less than 70 percent if even 30 percent of their original personnel were to deploy to an OOTW. If less than 30 percent of the unit deployed, we counted as involved in the operation only the actual number of people who deployed. Figure 5.3 illustrates how Figure 5.2 would change if we had omitted the effect of declaring the entire unit to be involved if 30 percent or more of it was deployed to Somalia. The cross-hatched area at the immediate top of the bars indicates the effect of counting whole units if over 30 percent of the personnel deployed as opposed to counting only the personnel who deployed. In our analyses we have included these effects.

Figure 5.3 depicts the total number of personnel in theater, the total number of personnel involved in the operation if we count only the people who deployed to Somalia as involved in the operation, and the total number of personnel involved if we count complete units as involved if they are over 30 percent deployed. Clearly, if we ignore the effect on units of deploying a significant fraction of their personnel, then the total number involved falls by an average of 22 percent. However, the overall shape of the curve is unchanged, with peaks in May and July.

The above analysis assumes that the TPFDD that we are using reflects the units that deployed personnel and were affected by the operation. We know that the TPFDD records understate the demand in

[1]Unit here refers to an "AA" level unit of organization, usually a separate company or battalion.

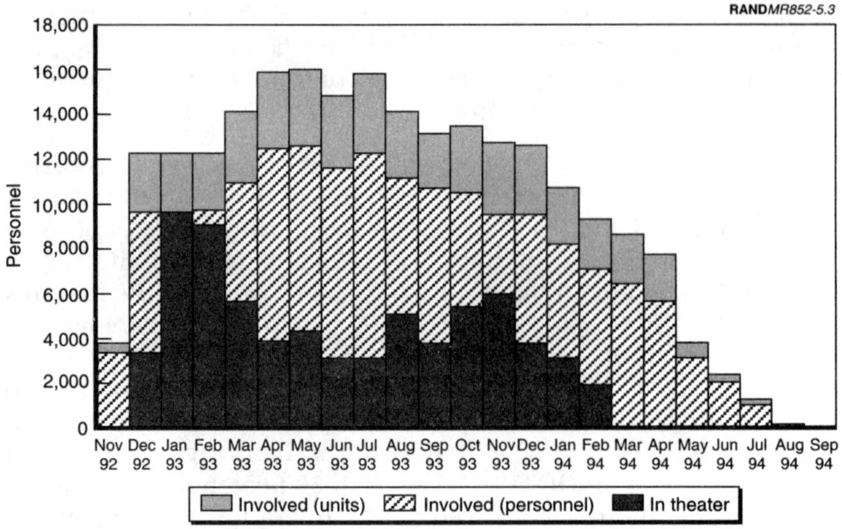

Figure 5.3—Effect of Counting Units

many ways. We know, for example, that many of the deployment records identified by the UIC were in fact either task force organizations or were composite units in the sense that the deployment record actually reflected the deployment of personnel from a number of different units. In addition, some personnel deployed to the theater who were either not on the TPFDD at all or deployed but were reflected in the TPFDD in what are known as nonunit personnel records. In the latter case the unit identifier is not contained in the TPFDD, and we have no way of determining it from the information in the TPFDD.

In assessing the impact of OOTW on the Army, we also conducted a limited analysis of personnel deployment information obtained from the Army personnel data system. In the case of Somalia, the personnel data indicated that personnel deployed from almost 1,500 different units (four-digit UIC level, which would normally indicate a battalion or separate company-size unit), only 20 of which deployed 100 or more persons. The TPFDD we used identified fewer than 250 units in deployment records. In the case of MP units, the TPFDD identified only ten MP companies. The personnel deployment data showed MPs, personnel in CMF 95, deployed from 62 different UICs. Forty-two UICs were of MP companies, and ten were of battalion headquarters detachments. The remaining ten were brigade headquarters, criminal investigation units, etc. We found similar results for deployments to Haiti. We will take up this point and its possible implications later in this discussion.

The CALL report also identified the problem of having a slice deployed and the effect on the units left behind. Currently the USR does not reflect the fact that people are deployed and therefore not readily available for deployment to an MRC or for collective training at home station. In some cases the missing personnel may severely impact the rest of the unit's ability to conduct useful training exercises.

EVALUATING THE EFFECT OF OOTW MISSIONS ON ARMY MRC CAPABILITY

Using the results of our MRC analysis (summarized in Chapter Two), the methodology discussed in the previous chapter, and the requirements for OOTW, we can evaluate how performing OOTW missions affects the Army's ability to respond to an MRC. Specific units are needed to support the early-deploying combat force and are designated as FSP units. The active component of the FSP is about 70,600. Our analysis of the number of support personnel required to deploy early in support of a four- to five division MRC indicated that, depending upon the theater and the timing of the early deployments, about 40,000 to 67,000 would need to come from the active component; the rest could come from the reserves, based upon expected reserve component readiness and assuming mobilization of the reserve occurred immediately (Sortor, 1995).[2]

[2]We recognize that the Army statement of requirements has been questioned and that alternative statements exist. In our previous report (Sortor, 1995) we discussed alternative support force requirements for similar SWA combat force levels that ranged from 136,000 to 180,000. These differences, however, do not necessarily translate into similar differences for the portion that must be supplied from the active component. Increasing the later-deploying force (reflecting a full doctrinal statement of requirement versus some minimum essential level of services) increases the reserve portion of the requirement but not the active portion. Changing the timing of the early-deploying force does significantly affect the active requirement. The timing of the early-deployment capability reflected in TAA01 as compared to the timing in the

Table 5.1 summarizes and compares the active component MRC requirements for a range of scenarios to the aggregate active component FSP capability and shows the portion of the force that would be available for OOTW under each scenario.

For the remainder of this analysis we will use the case where the active component would need to supply 59,000 support force personnel as reflected in the SWA 2 scenario case. This is the case where mobilization occurs simultaneously with the initial deployment of forces and the capability exists to deploy the support forces at the timeline desired by the Army and reflected in its force planning for TAA01.

Figure 5.4 depicts the number of support personnel involved in Somalia, almost all of whom came from the active component high-priority CONUS-based units needed to support an MRC. The line depicts the number of active component personnel in the FSP who are not required to deploy early in support of a four- to five-division MRC, provided reserve mobilization occurs immediately. Another case, NEA in Table 5.1, is the NEA case where it is the first contin-

Table 5.1

Active Component Support Force Requirements Versus Capability

Scenario	AC Required for Scenario (M = C)	AC in FSP	AC Available for OOTW
SWA 1	37,000	70,600	22,400
SWA 2	59,000	70,600	10,400
NEA	67,000	70,600	2,400
Nearly simultaneous (SWA and NEA)	59,000 to 67,000	70,600	2,400 to 10,400

NOTE: See Sortor (1995) for a discussion of the alternative scenarios and requirements.

---

example discussed in Chapter Two increases the active requirement from about 37,000 to 59,000. Most of the increase is in transportation requirements. With the current and the planned force structure, the active component cannot supply the number of transportation units required, so either the reserve component units will need to deploy more quickly than our model would predict or the theater will be short of the desired capability until the reserve component units can be mobilized and deployed.

gency. In the latter case there are not enough personnel even if we assume the transportation, train-up, and retraining time is zero. In these cases, for the least demanding deployment schedule for SWA (SWA 1 in Table 5.1) there are 22,400 active component personnel not required, there are 2,400 for NEA, and there are 10,400 for SWA with the Army preferred deployment schedule (SWA 2 in Table 5.1).

Figure 5.4 shows that there are not sufficient active component personnel in the FSP to provide the number of support personnel required by an OOTW of Somalia's magnitude and to meet the requirements for deploying fully ready and capable units early to a single four- to five-division MRC. In addition, the active component in the FSP may be called upon to support more than one OOTW at any one time. For example, a total of 25,000 active Army personnel were deployed to Florida following Hurricane Andrew, which occurred only a few months before the deployment to Somalia began. If Andrew had struck while the deployment to Somalia was under way and half of the 25,000 active Army personnel had been support personnel from the FSP, which includes almost all of the CONUS-based active component support personnel, then the line on Figure

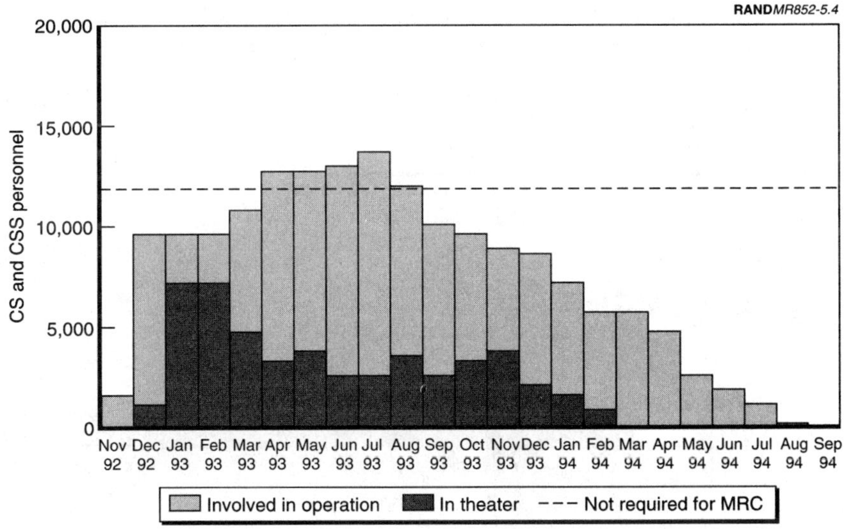

Figure 5.4—Somalia's Impact on the FSP

5.4 would have indicated even fewer combat-ready active component personnel in the FSP able to deploy early to an MRC. Further, forces were deployed to Rwanda, Panama, MFO-Sinai, Macedonia, and Guantanamo Bay while forces were involved in Somalia. The above also does not take into consideration the specific units involved. While there may be sufficient or even excess personnel in some branches, others may be severely short. Recall from our earlier discussion that even in the least demanding MRC case, our earlier analysis showed there would be a shortage of active component units in ordnance, quartermaster, and transportation. We look next at particular branches and type of unit.

Our previous analysis found that several branches would have insufficient personnel to meet the requirement for early-deploying support personnel for a single four- to five-division MRC even in the absence of an ongoing OOTW. Figure 5.5 shows the results of our earlier MRC analysis, along with the current branch distribution for active component personnel in the FSP.

Figure 5.5 compares the active component content of the FSP (the dark bars) with the requirement for active component units under

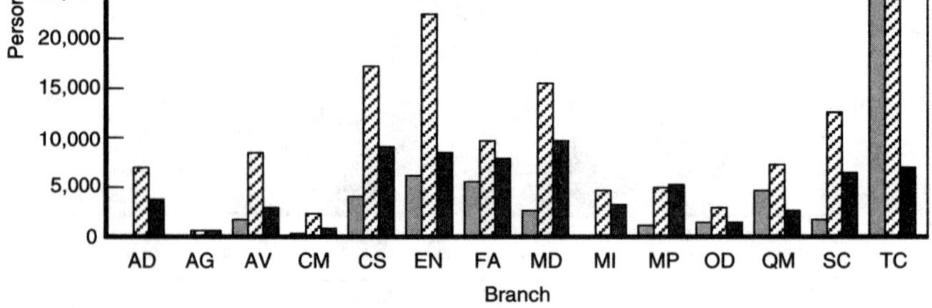

Figure 5.5—Branch Shortfalls in Active Component Units

two mobilization assumptions. Only the quartermaster and transportation branches indicate a significant shortfall in meeting the active component requirement if mobilization is declared at C-day. If mobilization is delayed for 20 days (M-day = C-day + 20) as was the case in ODS, only the military police would have roughly the number of forces required, while all the other branches would be short.

In Figure 5.6 we add the effect of Somalia to the active component MRC requirement shown in Figure 5.5 (the cross-hatched bars) and compare the result to the active component FSP capability (dark bars). Most branches have sufficient active component personnel to meet the requirement to deploy early to an MRC (assuming immediate mobilization of the reserve components). Only quartermaster and transportation would be short of personnel overall in the branch. The shortfall increases when the forces involved in Somalia are added to the MRC requirement. These overall comparisons, however, mask shortfalls in particular types of units that are typically required for both OOTW and the early deployment to an MRC.

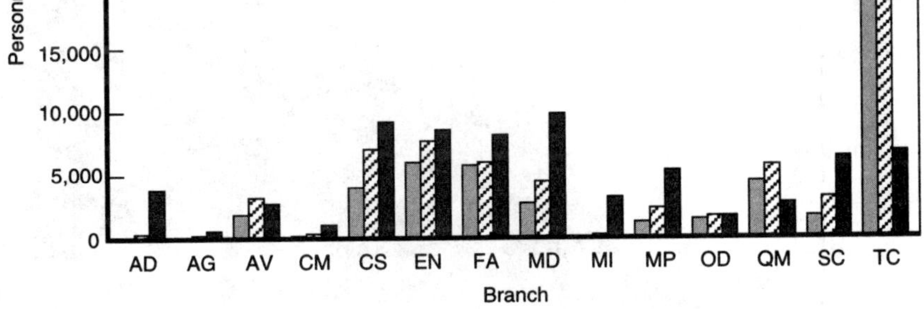

Figure 5.6—Comparison of Active Component Requirements and Active Component in FSP

Recent deployments, including Somalia, have stressed some parts of the force more than others. As we noted in an earlier chapter, different types of OOTW require varying mixes of unit types and also varying numbers of units and personnel. In some cases there simply are not many units of certain types in the active force structure, and reserve component units may not be available. For example, reserve component units were not generally available for Operation Restore Hope in Somalia. There are four supply companies in the active component in the FSP. In Restore Hope, two active component supply companies were deployed during the first rotation. They were replaced by one and a half supply companies in the second rotation and a single supply company in the third rotation. Our previous analysis of one modest-sized MRC indicated a requirement for twelve of these supply companies and that at least two must come from the active component if they are to reach the theater at the prescribed time and at the required readiness level (Sortor, 1995). These MRC requirements are depicted in Figure 5.7.

Besides the four companies in the active component of the FSP, there are six additional companies in the reserve component of the FSP. In addition to units in the FSP, there are six more units in the

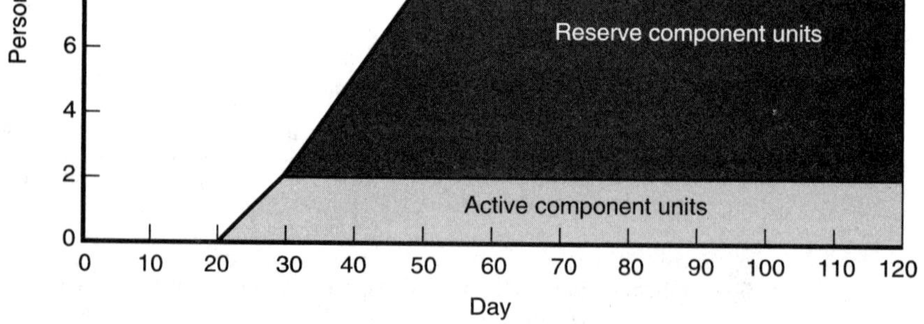

Figure 5.7—MRC Supply Company Requirements and Mix

active component and sixteen more in the reserve component force structure. So if all supply units could instantaneously be available for OOTW or MRCs while at home station, the force size and mix would be more than adequate for this situation. There could be two active component units in Somalia and two in the CONUS ready to deploy to an MRC, assuming that no other OOTW requiring supply companies was also under way.

But these units are not instantaneously available. The CALL data discussed earlier indicate that CSS units in general would require at least a couple of weeks to prepare for deployment and another 10 to 28 weeks to recover after returning to home station. This would total about three to eight months, in addition to transit time, that these units would be unavailable for deployment to an MRC. For illustrative purposes let us assume preparation and recovery time for these supply companies is a total of only three months (one to prepare and deploy to the OOTW and two to recover and regain their ready status for an MRC). Under this assumption, the number of supply companies involved over time in Operation Restore Hope is shown in Figure 5.8, together with a line representing the number of active component units in the FSP that are not required to deploy early in response to an MRC. As noted earlier, it is assumed that units are not available for early deployment to an MRC if they are preparing for an OOTW deployment or recovering after such a deployment.

In this case there are just sufficient units in the active component to simultaneously support the OOTW and maintain a sufficient force to meet the MRC requirement for the first four months of the operation. However, because of the need to rotate the units and the time needed to prepare for deployment, there are not sufficient units once the units for the second rotation begin preparing for an OOTW. Not until November after the OOTW requirement decreases does the active component force have the capability to simultaneously support its OOTW and MRC requirements. If the force requirement had remained steady at two supply companies, then the shortfall in ready units able to respond to an MRC would have continued to exist.

It should be noted that there are six active units of this type that are not in the FSP (all are stationed overseas). One of these units deployed to Somalia, as part of the initial deployment. The other four of the five units that deployed were the four active component FSP

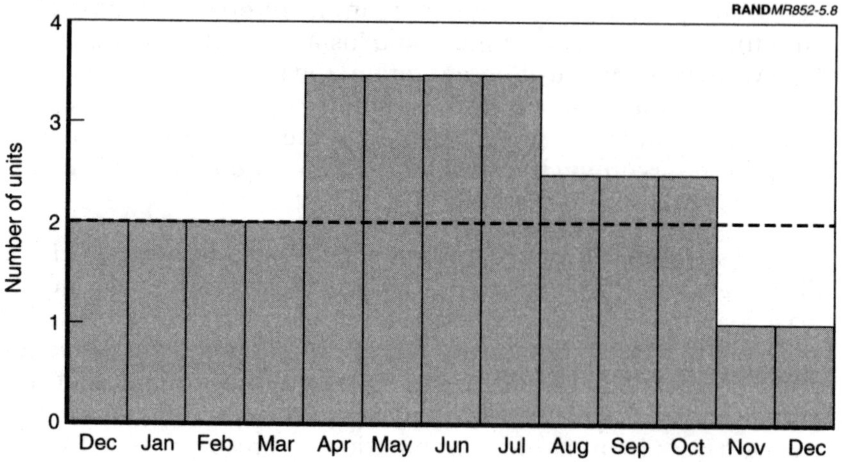

Figure 5.8—OOTW Supply Company Requirements

units. For our baseline case we do not consider the forward-stationed units available for out-of-theater OOTW, though they could be used in some circumstances as they were in this instance.

One unit, the 364th Supply Company, also illustrates the necessity for some units to be repeatedly deployed on OOTW missions. The 364th deployed a small party (18) as part of the initial deployment to Somalia (this has been ignored above). The entire unit (145 personnel) deployed as part of the third deployment. They deployed in August 1993 and returned to the CONUS in January 1994. This same unit deployed (130 to 140 personnel) to Haiti in September 1994.

Another example illustrates the potential importance of the preparation cycle for unit availability. Analysis of the MRC requirements indicates that there is a requirement for 16 combat support equipment engineering companies. Only one of them needs to come from the active component to meet the readiness criterion for the MRC. The active component of the FSP includes five such units, while the reserve component of the FSP contains seven units. At the time of the Somalia deployment only four active units were designated for the

CFP.[3] All four of the active component units in the CFP deployed to Somalia. One of these units (642 EN CO), which was in the initial deployment to Somalia, was also in the initial deployment to Haiti. There were also three active component units of this type that are not in the CFP. None of these units deployed to Somalia or Haiti. Two of the units were stationed overseas, while one was at Fort Hood, Texas. For this discussion we will consider the units in the CFP as it existed at the time. We recognize that an additional active component unit has now been added to the FSP. Thus there were, by this calculus, an adequate number of ready units to meet the MRC requirement.

Now consider how the deployment to Somalia for Restore Hope influenced the availability of early-deploying active units. Once again, we use the actual deployment timeline for Restore Hope. Also, it is assumed that units are not available for early deployment to an MRC if they are preparing for and deploying to an OOTW deployment, assumed again to take one month, or returning from and recovering after such a deployment, also again assumed to take two months.

Figure 5.9 illustrates the number of active component units involved in the OOTW (including preparation and recovery) over time. It also indicates the number of active component units in the CFP that would not have been required to deploy early in response to an MRC.

In this case there were sufficient units in the force structure to both satisfy OOTW requirements and also maintain sufficient ready units to meet the MRC requirement for early-deploying units. However, the CFP included only just enough units to satisfy the MRC requirement before meeting the requirement of any OOTW. Therefore, Figure 5.9 also indicates a potential shortfall of from one to three units in the MRC requirement for later-deploying units; this would have had to be met by withdrawing units from the OOTW or drawing upon the active component and reserve component units outside the CFP.

---

[3]At the time of Somalia, the high-priority support units designated to support the first MRC were identified as CFP (Contingency Force Pool) units. The Army has since adopted a somewhat different concept for managing these units and has designated them as FSP (force support package) units.

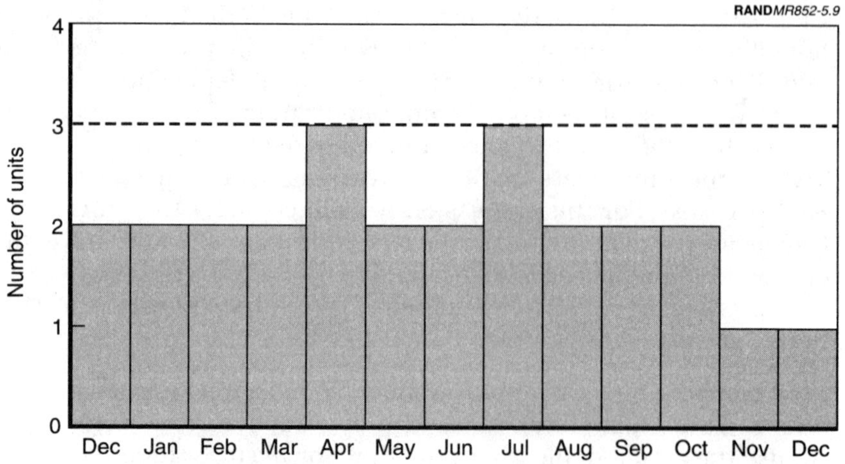

Figure 5.9—Engineer Combat Support Equipment Company in OOTW

If we change the assumptions about how long it takes a unit to prepare and recover from a deployment, then the above graph will change markedly. Figure 5.10 shows the effect of lengthening both the preparation time and the time to regain ready status by one month each. That is, assume it took two months to prepare and deploy each unit and three months to recover and retrain them. The graph would also change markedly if we altered the length of the deployment and, when multiple units are to be rotated, the phasing of the unit rotation.

Note that in this case, during five of the months the OOTW requirement is increased by one unit over the previous case. While there is still a sufficient number of units to satisfy both the OOTW and MRC commitments, it is obvious that the train-up cycle time could severely impact the availability of some types of units. And if this type of unit was required simultaneously in a second OOTW—and engineers are involved in many types of OOTW—then the number of units would not be sufficient to meet all mission requirements.

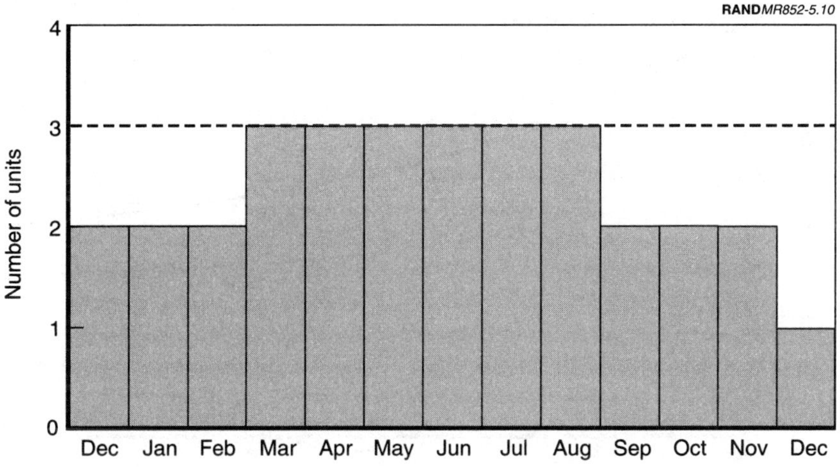

Figure 5.10—Effect of Lengthened Training Cycles

EFFECTS OF ALTERNATIVE FORCE ROTATION POLICIES

The above illustrations took the actual deployment schedules and thus rotation policy as a given. As we discussed earlier, the interaction between the time it takes units to prepare for and recover from a deployment and the policy followed for rotating units in and out of the operation can have a major influence on the number of units that would be available for an MRC. If units are simply placed in the operation for its duration, then the number of units unavailable for an MRC will equal the number deployed. On the other hand, if units are rotated frequently, say every 60 or 90 days, and units require considerable time to prepare and/or recover, then the number of unavailable units can increase markedly. Figure 5.11 shows the relationship between the number of units involved in a long-running OOTW, the tour length, and the time taken to prepare, transit, and recover (PTR).

Figure 5.11 depicts the case where ten units are deployed at any one time on a long-running OOTW. Notice that the number of units involved in the operation—preparing, transiting, participating, or

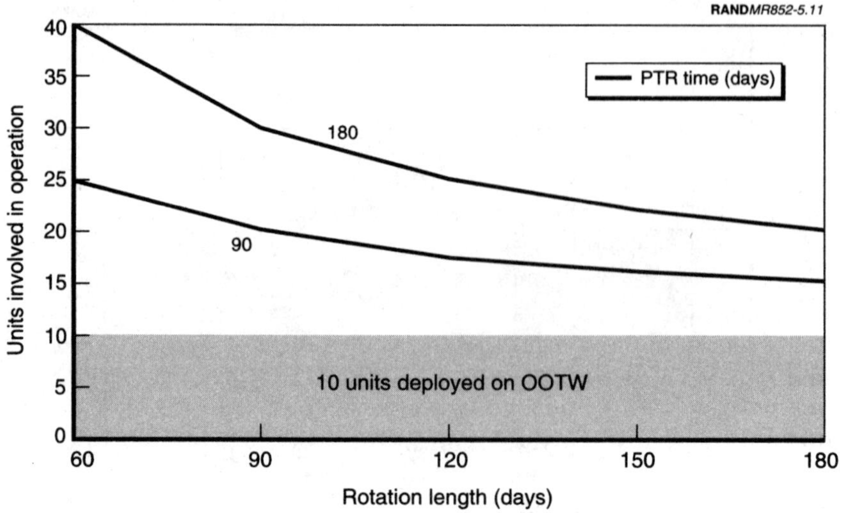

Figure 5.11—Effect of Varying Tour Length and Prepare, Transit, and Recover (PTR) Time

recovering—decreases as the tour length is increased, assuming no change in the length of time taken to prepare, transit, and recover. Also, the number of units involved in the operation increases as the length of time taken to prepare, transit, and recover increases, assuming no change in the tour length. In the real world it is likely that these two times will be related. In particular, as the tour length is increased, then the time taken to recover to full combat readiness is liable to increase. However, the transit time is not likely to change if the tour length is altered and will depend much more on the location of the OOTW and on the mode of transport used. Thus the interaction of these times is liable to be complex but readily computed based on policy options of interest.

There are 44 MP combat support companies in the active component. Let us take the case where 16 of these units are required to be ready to deploy early to an MRC and therefore must come from the active component. If more than 28 units are involved in the OOTW—in terms of being engaged in the theater, preparing for deployment, or still in the process of recovering to readiness status and therefore not fully combat ready—there is a potential problem. Further let us assume that 10 units are required to be deployed on a long-running OOTW, or set of operations, at any one time. This case is illustrated in Figure 5.12.

Figure 5.12 shows that in the case discussed above, for less than 28 units to be involved and therefore leave 16 units combat ready, the tour length must be set at a minimum of 120 days if the expected PTR time is 150 days. If the preparation, transportation, and recovery time is expected to be no more than 90 days, then the tour length can be as little as 60 days. Diagrams like this can be constructed for various combinations of tour length and preparation, transportation, and recovery time in order to determine a feasible tour length policy for a given OOTW that will ensure sufficient units ready to deploy to any potential MRC. Recognize also that the retraining time can be affected by policy. Giving greater priority to recover and refurbish equipment and for training resources for returning units can significantly reduce retraining time.

When we are designing a future force, we can use the same methodology to identify the number of units needed in the force structure to

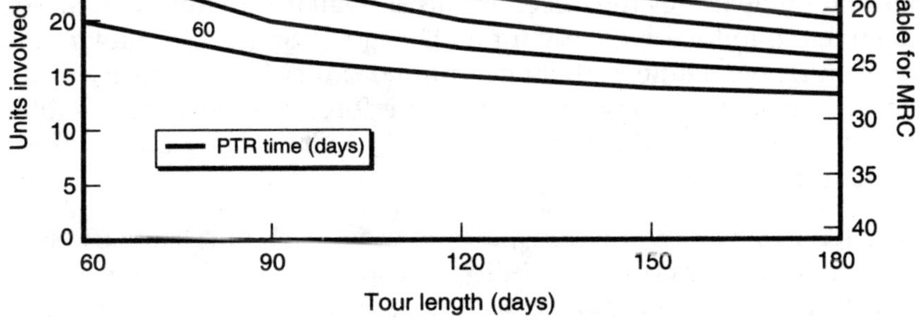

Figure 5.12—Example of Force Availability

ensure that planned OOTW requirements can be met and that a sufficient number of units will always be ready to deploy to a major regional contingency should it be necessary. The force structure can be determined based on a desired tour length and rotation schedule defined in terms of how long units are to be deployed to an OOTW and on the desired time between deployments. At the time of execution of an OOTW, however, the number of units that exist is fixed and only the tour length and, possibly, the recovery time can be affected. The Army can decide, for example, to leave units in the theater longer in order to reduce the number of units that are unavailable for an MRC. This would have morale implications, depending upon the type and length of the operation. Note that this is the policy being followed in Bosnia, where most active Army forces were deployed with the plan and expectation of staying for the stated duration of the operation of one year. In other recent OOTW deployments, however, Army units have typically rotated after 90 to 180 days.

Another option after setting the rotation policy is to give returning units greater priority for resources and thus accelerate the recovery time. In addition, units could be identified prior to their return and the recovery process altered to mirror that discussed earlier for units identified for immediate deployment to an MRC. This would also have personnel morale implications and would probably not be possible on a routine basis. Note also that once a unit returns and begins the process of block leave and personnel restructuring, it would be almost impossible to reverse this course and move from the return-to-readiness timeline to the return-to-readiness-for-an-MRC timeline.

Figure 5.13 illustrates the effect on forces available of applying more resources to reduce recovery time. Using the Somalia deployment and rotation schedule, reducing assumed transit and recovery time from 150 days to 60 days would cut the force that is not available during June, July, and August by roughly 4,000 and by almost that much through December.

Although past OOTW missions, for the most part, have not stressed Army forces in terms of their ability to respond to our baseline MRC, we did find cases where there would be insufficient active component units to simultaneously engage in an OOTW scenario and maintain sufficient units at a readiness status desired for deployment to

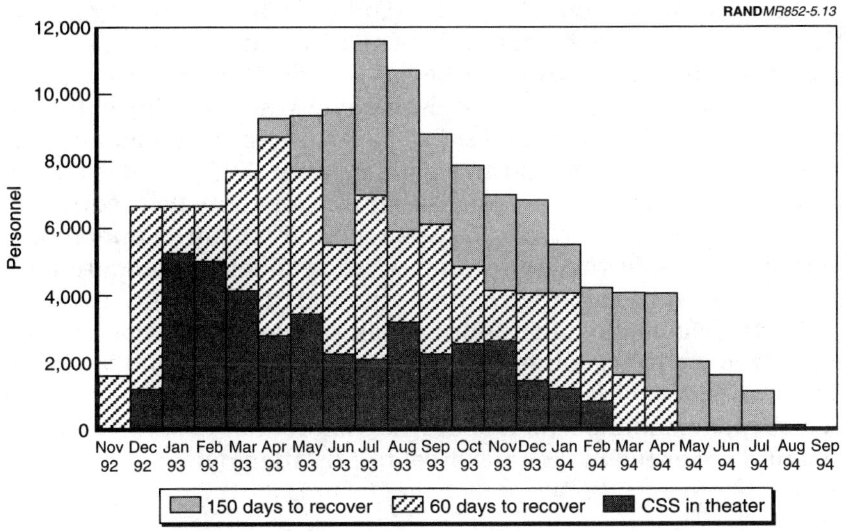

Figure 5.13—Effect of Accelerated Recovery Time

an MRC. For example, in six instances (general supply companies, air terminal movement control detachments, medium truck–POL, cargo transfer companies, water purification (ROWPU) detachments, and perishable subsistence teams) all of the active Army units planned for early deployment to an MRC were used in Somalia. Units were not all deployed at the same time but were rotated. At any point in time only one unit was in Somalia, with another unit getting ready to deploy and possibly another unit undergoing reconstitution. If an MRC had occurred, the Army would have stopped rotating units and/or would have needed to move these units immediately (if that were possible), substitute similar active component units (where they existed and were available), or prepare reserve component units for deployment more quickly than planned.

Future OOTW requirements may call for additional units and stress the active forces to the point of rendering them unable to respond on short notice to an MRC. Clearly, responding to multiple OOTW simultaneously could require more of many types of units than exist in the active force structure. During part of 1994, for example, the Army was simultaneously engaged in operations in Macedonia (Able Sentry), Cuba, Sinai, Rwanda, Haiti (Uphold Democracy), Kuwait (Vigilant Warrior), fire fighting in the CONUS, and a number of lesser operations around the world. During this period the Army would not have been able to respond to a fast-breaking MRC with the preferred mix of ready forces. Two suggestions have been raised for addressing this problem.[4] One is to add active units of certain types to the force structure. This solution presumes that we can correctly predict the unit types and numbers that may be required for future operations and that sufficient budget resources are available to procure and maintain the units in the active structure. Clearly, however, this is an option to be considered and for a small number of critical units is probably the correct solution. The second suggested solution is to make greater use of existing reserve component units in OOTW. The next chapter examines the role of reserve forces in OOTW both in a historical sense and in their potential role for the future.

---

[4]A third alternative is to make greater use of capabilities from outside the Army. The other services could provide many of the same types of capability, as could the private sector through contract. For example, the Air Force provided engineers for construction at the Tuzla air base in support of Operation Joint Endeavor, and Brown and Root provided support services in Somalia, Haiti, and Bosnia. We have not examined the extent to which these alternative sources of services might be expanded to further substitute for Army force structure in the future.

Chapter Six

ROLE OF ARMY RESERVES IN OOTW

INTRODUCTION

For most OOTW contingencies in the past, the active component has provided the bulk of the forces employed. Until recently, the reserves were not usually considered in initial planning for OOTW contingencies, as the assumption was that they would not be available. The BUR, however, called for an increased role for the reserve component in both war and peace operations. Even with this increased emphasis and planning guidance from OSD, military force structure and operational planners have been reluctant to rely on the reserve component. This is not to say that reserve component forces and personnel have not been participants in OOTW. They have participated to some extent in many.

As will be discussed further below, a USACAA study documented Army reserve component participation in 17 of 49 operations conducted between 1975 and 1990 (USACAA, 1991). Participation was usually in volunteer status or during the unit's regularly scheduled two- or three-week annual training period. The presumption on the part of most planners, especially joint planners, was that the political will would not permit the involuntary mobilization of reserve component soldiers to serve in OOTW where U.S. vital interests were not clearly threatened. This meant that only volunteers were potentially available. The Army has not considered volunteers adequate for many of its missions, since it wanted whole units and could not rely on getting whole units or even large portions of a unit. The Air Force, however, has relied on individual reserves in a voluntary status for many of its missions and utilizes them on a regular basis for both normal peacetime operations and OOTW.

In recent operations in Somalia, Haiti, and Bosnia, the Army has used volunteers as well as involuntary recall under the Presidential Selective Reserve Call-up (PSRC) authority in Section 673b of the U.S. Code.[1] The rest of this chapter will describe the historical role of the Army reserves in OOTW, how this role is changing, and what options may exist for the future.

HISTORICAL ROLE OF ARMY RESERVES IN OOTW

Army reserves, both Army National Guard (ARNG) and U.S. Army Reserve (USAR) personnel, have participated in numerous OOTW over the years. For example, USACAA (1991) documented that between 1975 and 1990, 17 operations out of the total of 49 involved reserve forces, with the reserves accounting for about 11 percent of the total effort (in terms of man-days). Table 6.1 summarizes the man-days for active versus reserve for each of the major operational categories used in the USACAA study.

What factors led to reserve participation in some operations and not in others? It is believed that certain characteristics of the operations and the circumstances of the time facilitated the use of reserves in some, while almost dictating, and certainly encouraging, the use of active component soldiers in others. The characteristics that greatly facilitate the use of reserves included: short operations that permitted the use of reserves in volunteer or annual training status, a relatively lengthy planning horizon, a firm concept of operations, and an operational environment that was fairly benign with respect to threat or chance of hostile actions against deployed U.S. forces.

---

[1]There are several statutory provisions for calling up reserve units; the most commonly referenced are the PSRC, "partial mobilization," and "full mobilization." Title 10 USC, Section 673b, provides authority, called PSRC, for the President to activate up to 200,000 members of the Selected Reserve for up to 270 days for an operational mission at times other than war or national emergency. The other provisions, which permit more extensive call-ups for up to an indefinite period of time, require a declaration of national emergency or war.

Table 6.1

Man-Days for Active and Reserve

Operational Category	Active Army	Reserves
Combat operations	506,024	0
Peacekeeping	2,812,589	0
Show of force	44,366	0
Security augmentation	914,909	113,075
Nation building and support	1,396,659	655,637
Humanitarian assistance	793,359	6,176
Disaster assistance	366,472	79,137
Support to law enforcement	252,511	3,062
Refugee resettlement	1,250,444	165,812
Total	8,337,333	1,022,909

For reservists to participate in military operations—whether an MRC, general war, or an OOTW—they must be brought to a full-time active status. This may be as volunteers or nonvolunteers and may be in federal status or, uniquely in the case of the National Guard, in state status. Participation in the operations summarized in Table 6.1 was as volunteers or, more frequently, in conjunction with a regularly scheduled annual training period. In contrast, units participated in Operation Desert Shield/Storm under three different authorities. Initially, some served as volunteers, followed by service under the President's authority using Title 10 USC 673b, and finally under the President's partial mobilization authority using Title 10 USC 673.[2] Each of these has both advantages and disadvantages and is more or less appropriate depending on the particular operation and circumstance.

In the case of the National Guard, the governor of the state may activate his National Guard units in state status for emergencies requiring augmentation of other state resources, for example, in the event

[2]See Sortor et al. (1993) for a more complete description of the mobilization actions for Operation Desert Shield/Storm.

of a riot to augment local and state police authorities.[3] The Army and the Air National Guard may also be federalized under Title 10, sections 3500 and 8500, to repel invasion, suppress insurrections, and enforce federal laws.

Army reservists (ARNG and USAR) may volunteer for active duty and serve on temporary tours of active duty (TTAD) to support Army operations. In the past, the lack of funds to pay reservists has been a major limitation on greater use of volunteers to support operational requirements. Reservists may also participate in operations as a by-product of training, as in the case of many of the operations in SOUTHCOM. Units deploy during their annual training period and participate in operations rather than performing their annual training at a CONUS training facility. The ARNG, for example, annually deploys almost 14,000 reservists to SOUTHCOM to perform engineering, medical, and security tasks. The USAR also deploys personnel on overseas deployment training events. These training periods both support the unit's training and meet OOTW mission needs.

Only recently, as we will discuss further in the context of recent OOTW, has use been made of the President's authority under Title 10 USC 673b (termed PSRC) to activate reservists for OOTW. In fact, the exercise of PSRC for Operation Desert Storm in 1990 was the first use of the authority since it was passed in 1976. Many military planners had considered PSRC as unavailable for OOTW or any operation short of an MRC. This lack of confidence in the ability to gain access to reservists when needed is one of the reasons military planners have been so reluctant to plan on the use of reservists in OOTW.

As a result of the military's reluctance to depend more heavily on use of reserves for OOTW, coupled with the desire on the part of OSD to encourage greater use, RAND conducted a study to identify impediments to the use of reserves and recommend actions to alleviate if not eliminate the impediments (Brown et al., 1997). The study formulated a framework for analyzing the influences on the use of reserve component versus active component forces and defined three sets of influences that were believed to be key—processes, players, and factors. The study then examined these influences and how they

---

[3]See Brown, Fedorochko, and Schank (1995) for a discussion of the state and federal roles of the National Guard.

affected the demand for forces, the selection of forces to meet the demand, and ultimately the decision to either use or not use reserve component forces in a given contingency.

The processes included the mission planning process for determining the demand for forces, the force selection process for identifying the specific sources of supply for the forces, and service cultures. The study concluded that service cultures play an important role and form the basis for integrating the supply to meet the demand for forces. The players were the unified commands that have responsibility for defining operational requirements, the force providers that decide what specific forces to provide, and the military departments that integrate the services' need for reserve forces. The factors were operational, resource, and institutional and were not unique to any one process or one player. They influence all aspects of specifying the demand and selecting forces, and they reflect the integrating effect of service cultures on the forces available to be selected and used in a given operation. It is these factors that are normally thought of as being the major influences on whether reserve component forces can be used or not.

Of the eight factors that commonly shape the demand for forces for OOTW (task-resource requirements; scope of the operation; urgency; duration of the operation; level of threat; level of control; treaty, policy, or mandate restrictions; and the involvement of nonmilitary organizations), three were identified that generally work against the use of the reserve component—urgency, duration, and level of threat. The others could work either way. Two major influences were identified that affect the forces supplied—institutional and resource—because they affect the criteria used in identifying and selecting forces. These criteria include availability, functional requirements, responsiveness, level of risk, perceived importance and national acceptance, and accessibility. Only the last, accessibility, is unique to the reserve components. Accessibility is normally thought of in a single dimension—has authority been given to use the reserve components or, if not, is it likely to be given if requested? The study points out another dimension that has affected the selection of forces in recent operations—is funding available or likely to be made available to pay for the use of the reserve components in a full-time capacity?

The above study was in terms of OOTW conducted outside the CONUS. (All twenty-three operations identified in the analysis were peacetime contingencies outside the CONUS.) Domestic missions, however, are also an important consideration in assessing the impact of OOTW on Army forces. In the case of domestic missions, the ARNG has the primary responsibility for augmenting state and local resources in responding to domestic emergencies (DoD, 1994). While the domestic missions are less demanding in many respects than contingency operations overseas, they do impact the availability of forces and, in the case of national disasters such as a major earthquake, may take priority over immediate response to what would otherwise be perceived as requiring immediate overseas deployment of U.S. forces. Further, some domestic missions (drug interdiction and counterdrug operations, for example) have increased somewhat dramatically. Suggestions have been made to increase the use of reserve forces for border protection as well.

Another recent RAND study examined the role of the ARNG in domestic missions and assessed the impact of such missions on the ARNG's federal role in responding to war or major combat contingencies (Brown, Fedorochko, and Schank, 1995). The study determined that the ARNG was seldom employed in state duty in large numbers for lengthy periods. It pointed out that fiscal year 1993 showed the highest level of state active duty in over a decade and involved slightly over 460,000 man-days by over 34,000 Army and Air Guard members. This equated to about 6 percent of the total available strength and is less than one duty day per member. The ARNG may also engage in domestic OOTW in a federal status. The USACAA report identified ten domestic operations involving either ARNG or USAR personnel. It did not identify which component provided the personnel.

As noted earlier, accessibility has been one of the concerns cited as an impediment to greater reliance on reserve component forces for future OOTW participation. The issue was discussed in a DoD report issued in April 1994 (DoD, 1994). The report emphasized the need for reserve component use in both domestic emergencies and peace operations in order to carry out the national military strategy. It defined the new DoD policy for use of the reserve components by categories of missions as follows:

- For major regional conflicts (MRCs) and major domestic emergencies, access to Reserve Component units and individuals through an order to active duty without their consent will be assumed.
- For lesser regional conflicts, lesser domestic emergencies, and peace operations where reserve component capabilities could be required, maximum consideration will be given to voluntary access to reserve component units and individuals before seeking an order to active duty without their consent.

Note the explicit reliance on volunteers to the maximum extent possible for OOTW contingencies other than major domestic emergencies. While the Air Force has been successful and comfortable with relying on volunteers for most if not all of its OOTW contingencies, the Army much prefers utilizing involuntary recall under PSRC authority. In some cases the Army has requested PSRC, either formally or informally, and been denied that authority. In the case of Just Cause, the CINC requested PSRC and it was not forwarded from the Joint Chiefs of Staff (JCS) to the Secretary of Defense. Volunteers were solicited and used, but the experience was not satisfactory because of delays in getting personnel, because some skills were not available, and because the short period that many volunteers were available created a lack of continuity that affected operations. There was also a lack of unit cohesion. For these reasons, many in the Army opposed the use of volunteers for OOTW.

In the case of Somalia, both SOCOM and Army Forces Command began the process for using PSRC and were told that the authority would not be requested from the President. Interviews with personnel at FORSCOM, Department of the Army, and the JCS indicated that the Army request and justification was not felt to be persuasive. This is explained in part by the fact that the mission changed during planning; the mission and objectives that existed at the time the plan was executed required fewer forces and thus were less persuasive in justifying PSRC. In the case of both Just Cause and Somalia, as will be discussed further below, the Army solicited volunteers and, in the case of a postal unit, formed a unit using volunteers from a number of sources for service in Somalia.

The General Accounting Office (GAO) in a recent report concluded that adequate numbers of reservists had volunteered when needed and had performed adequately (GAO, 1996). It also pointed out the differences between the services in terms of past reliance on volunteers and on the Army's preference for using PSRC to access reservists for OOTW. A number of recent operations have made use of both volunteers and involuntary recall and even a combination of the two. These will be discussed further in the context of recent operations in the Sinai, Somalia, Haiti, and Bosnia.

THE CHANGING ROLE OF ARMY RESERVES IN OOTW

Recently we have seen a greater use of Army reservists in OOTW in both voluntary and involuntary status. The use of PSRC authority to involuntarily call ARNG and USAR units and individuals to active duty has facilitated the expanded role of reserves in recent OOTW. We will discuss the role of the reserves in four recent operations: MFO-Sinai, Restore Hope in Somalia, Restore Democracy in Haiti, and Joint Endeavor in Bosnia. These cases illustrate three different modes or roles for the reserves. The use of contractors to varying extents in all cases also affected the need for and use of reserves. When contractors were used, fewer reserve forces units and personnel were needed to provide CSS functions.

In the case of the Sinai, volunteer reservists were sought for an active duty tour as part of a composite unit manned with a mix of active and reserve component personnel and formed expressly for this one operation. In the case of Somalia, while early consideration was given to using PSRC authority, volunteers were solicited for duty in a provisional reserve unit and smaller elements without resorting to PSRC to activate the personnel. In Haiti and Bosnia, reserve component units were involuntarily called to active duty using PSRC authority. As will be discussed below, however, units were sometimes selected for call-up on the basis of being able to get sufficient "volunteers" from the unit, as the guidance was to rely on the use of volunteers to the maximum extent possible. In some cases the units were used directly in the operation, and in others, particularly in support of Bosnia, the reserve unit was used to backfill for an active unit and to carry out that unit's normal peacetime duties in either the CONUS or in Europe. Each of these cases and the lessons learned are discussed further below.

MFO-Sinai

The Army has been rotating a battalion-sized infantry unit (the actual strength has been as high as 1,200 but more recently has been on the order of 1,000 soldiers) to the Sinai as part of the Multi-National Force and Observers Mission, Sinai (MFO-Sinai) since 1982.[4] (See GAO/NSIAD-95-113, 1995, for a description of the depth of involvement and costs for the U.S. participation in MFO-Sinai.) Until 1995 this U.S. force was drawn from Army active component units of the XVIIIth Airborne Corps.

In 1993 the Chief of Staff of the Army requested that the Army explore the possibility of using reserve component personnel for this mission. As a result, the six-month rotation from January 1995 to June 1995 was conducted with a unit that consisted of 20 percent active component soldiers, 70 percent soldiers from the ARNG, and 10 percent soldiers from the USAR.[5] This was a composite unit formed and trained explicitly for this one rotation with the intent that after the rotation the unit would be disestablished and the reserve component personnel returned to reserve status.[6]

The unit, a battalion task force, formed at Fort Bragg on October 2, 1994, and trained as a unit. The unit—designated the 4th Battalion, 505th Parachute Infantry Regiment—deployed to the Sinai in January 1995 and redeployed to Fort Bragg in July 1995. All of the reserve component personnel were to be released from active duty and returned to their civilian status by August 4, 1995. All of the reserve component personnel were volunteers serving under orders to extended active duty. The period of service was clear from the beginning, as was the location and purpose of service. This contrasted

[4]In addition, there is a multifunctional support battalion permanently stationed at the MFO-Sinai North Camp and manned with active component soldiers on one-year tours.

[5]See the ARI report, *Reserve Component Soldiers as Peacekeepers*, for a detailed analysis (Phelps, September 1996).

[6]See Brinkerhoff (1995b) for a more complete description of the use of volunteers for this mission.

with the case of MP units called to duty in support of Operation Maintain Democracy, in which this was not the case. Those units will be discussed in the context of the section on Haiti, below.

Somalia

Only one reserve component unit participated as a unit in Operation Restore Hope in Somalia. An Army Reserve unit, the 711th AG Unit (GS Postal), was formed as a provisional volunteer unit and provided postal support to Army personnel deployed to Somalia. Other reserve component personnel participated both in support of other units that deployed and in Somalia as part of smaller detachments or teams. The latter primarily provided training and equipment preparation support. Civil affairs reservists participated in Somalia, as they have in almost every recent OOTW. The SOCOM request for PSRC was prompted by the need for access to the reserve civil affairs personnel. The request, as discussed earlier, was not approved, and all personnel who participated were volunteers and served without being ordered involuntarily to active duty. PSRC was also requested by FORSCOM during the earliest planning phase for the operation, but the purpose and nature of the mission subsequently changed. This made the need for PSRC less clear, since many fewer units were required, and the request was not sent forward for presidential decision.

Somalia was supported by the Army using unit rotations that ranged from four to six months. This rotation policy, as compared to placing units in the theater for the duration of the operation or rotating individual soldiers, was followed for all units deploying to Somalia, including the postal units. At the time, there was only one active component postal unit in the CONUS available for deployment to a contingency operation. It deployed early and was the first postal unit in Somalia. The 711th was the second to deploy. Subsequent rotations were met using provisional units formed with active component personnel.

The experience with the 711th, while meeting the mission needs, was not satisfactory to many in the Army, both active component and reserve component. The provisional units, especially if formed with reserve component volunteers from existing units, were thought to be less capable than regular units, either active component or reserve component, and it was believed that the use of reserve component volunteers might weaken the case for involuntary recall of units in future operations. There was also concern that taking volunteers out of existing units, particularly if done on a more widespread basis, would reduce the readiness and capability of those units and render them unavailable should they be needed for other operations. Some of the problems with the 711th, however, were more likely a function of the fact that it was the first provisional postal unit formed for Somalia. Also, many thought that a reserve component postal unit had volunteered as a unit and therefore would deploy as a regular unit.

In addition to civil affairs personnel and the USAR postal unit, other reserve component elements also supported Operation Restore Hope. Two contingents of ARNG personnel prepared aircraft for deployment. The first, twenty-two soldiers from CTARNG and MDARNG, shrink-wrapped aircraft for deployment to Somalia, while the second—thirteen soldiers from CAARNG, CTARNG, MDARNG, MOARNG, and MSARNG—deployed to Somalia to prepare aircraft for redeployment. The ARNG also supplied training teams that deployed to Somalia to provide M60A3 crew and maintenance training and AH-1S Cobra transition training for Pakistani pilots and maintenance personnel. In all cases the reserve component personnel served voluntarily on active duty without benefit of PSRC authority.

Haiti

During operation Maintain Democracy, three ARNG MP companies were called to backfill CONUS MP units at Forts Drum, Bragg, and Polk. The active units had deployed to Haiti and had left the installations without adequate military police support for the garrison. In this case also the guidance was to seek volunteers to the maximum extent possible; however, the intent was that after units were found where adequate numbers of personnel would volunteer for active duty, the units would be activated using PSRC. The ARNG had previously designated some units as Humanitarian Support Units. In these units, personnel had volunteered to be ordered to active duty for humanitarian missions and had agreed to serve without the need for involuntary mobilization authority. The activation of these units, however, did not follow the process or assumptions under which the soldiers had agreed to serve.

Other units and personnel also served in support of Maintain Democracy, with some deploying to Haiti. Civil affairs and psychological operations personnel from the USAR and special forces companies from the ARNG are examples. The ARNG special forces companies deployed to Haiti to relieve active component Special Forces units and were planned to serve a 179-day rotation with ARNG units serving two rotations. In all cases the units were called to service using PSRC.

Bosnia

Reserve component units have been called to active duty using PSRC authority for Operation Joint Endeavor. Units were planned to serve for up to 180 days, with subsequent activations to replace units withdrawn and demobilized. While some units deployed to Bosnia, civil affairs and psychological operations personnel in particular, other units were deployed to Europe to replace active units forward deployed in Bosnia. In this case, initial planning called for the use of reserve component units, and PSRC was assumed from the beginning. This operation's use of reserve component units and personnel closely mirrors the use of the reserve components in MRCs.

Operation Joint Endeavor was still ongoing as this analysis was being completed, and any lessons to be learned are not reflected in our analysis. It does serve as a benchmark, however, and indicates that for future large-scale OOTW the reserve components may well be made available from the beginning. This would help eliminate the planning confusion and indecision that has resulted in the past when the call was to use volunteers to the maximum extent possible and it was unclear as to what authority or what time limit or limit on numbers to be called might be established. Such uncertainty meant that Army planners didn't know how much they could rely on the use of the reserve components, and it may have caused them to look for other potentially less effective options, such as the use of ad hoc active units or of very limited active capability where the force was already stretched more than might be desirable. The problems of excessive personnel tempo and repeated deployments of some units discussed earlier may have been at least partially the result of such confusion.

Other Recent OOTW Involving Reserve Soldiers

In addition to the larger operations discussed above, reserve component units and personnel have served in numerous other OOTW. The above are presented to illustrate the various authorities and circumstances of service that have occurred in a relatively short period of time. History provides many other examples, however, as we noted in the introduction to this chapter.

Army reserve component units and personnel participate in a number of other operations in SOUTHCOM and East European countries, in counterdrug operations in the CONUS, and in disaster relief operations in the CONUS and overseas. They have performed riot duty (in Los Angeles, for example), fought forest fires, and provided other support to local, state, and federal authorities.

The ARNG and the USAR have participated for many years in humanitarian assistance and host nation support missions in SOUTHCOM, among other places. The largest, Fuertes Caminos, includes road repair and expansion, water well drilling, airport runway repair, and road and bridge construction. Such operations are normally accomplished with three-week deployments during the unit's annual training period and thus do not require any additional mobilization authority or funding for personnel costs. Operations that can be scheduled far in advance and can take advantage of scheduled training periods are easily accommodated by reserve component units and at the same time offer interesting and useful training opportunities. Recently, Army reservists have begun participating in operations in Eastern Europe as well.

The ARNG is a major participant in counterdrug support to state and federal authorities and supply- and demand-reduction activities. For example, in fiscal years 1990 through 1994, ARNG personnel participated in 5,000–6,000 operations per year, with from nearly 3,000 to almost 5,000 ARNG personnel on duty each month in support of the counterdrug operations with Joint Task Force Six in the southwest border states (Posehn, 1994).

In operations like counterdrug and support to local authorities, the personnel serve in a state status (applies only to the ARNG), in active duty status for annual training, or in a voluntary federal status on TTAD. Only very rarely are reserve component units called to duty involuntarily for federal duty in these types of operations. Even in these latter cases, however, it is usually for only a very short time.[7]

EFFECT OF OOTW ON RC FORCE AVAILABILITY FOR MRCs

Thus far, reserve component participation in OOTW has been accomplished without affecting the ability of the ARNG and USAR to respond to an MRC with planned forces. The force is sufficiently large and the operations have been sufficiently limited that, in terms of the force as a whole, there are enough units remaining to take care of an MRC. This does not necessarily mean the remaining units are sufficiently ready, however, as many of the support units in the Army reserve structure are not resourced at the level required to maintain adequate readiness for early deployment. Only the units designated for the FSP are resourced at a level consistent with readiness for rapid deployment. While in most cases both the participation and performance of reservists has been satisfactory, there have been cases where the Army has not been satisfied. Most of these latter instances involved a reliance on volunteers to satisfy requirements for units (as in the case of the postal unit to Somalia) or reservists whose participation was so short that it caused problems in the continuity of operations.

Short-duration deployments have been a problem for civil affairs and psychological operations personnel, for example, where it takes time to win the support and cooperation of host-country civilians and their own military commanders. Individual volunteers often are not able to volunteer for long periods, and the short rotations that result do not permit effective interactions to develop. The Army also considers reliance on volunteers unsatisfactory if units are needed. Ad hoc composite units formed using individuals, particularly reservists, and deployed quickly do not have time to develop the unit cohesiveness and collective skills desired in most Army units. These concerns are not unique to the U.S. Army. Pinch (1994), in a study of Canadian participation in peacekeeping operations, found the following:

---

[7]See Brown, Fedorochko, and Schank (1995) for a more extensive discussion of the state and local missions and their demands on military forces, particularly the ARNG while in state status.

> Formed units have been preferred over ad hoc units because of the former's superior cohesiveness. Regular force units have been favored over reserve units, for reasons of personnel availability, predictability and the purported value of in-depth military experience in performance . . . .

Obtaining sufficient volunteers from individual units so that they could be called as a unit has not been successful in many cases. The postal unit deployed to Somalia in 1992 was not able to obtain 49 volunteers from a single unit and had to draw from a number of units. The ARNG had a similar experience when it tried to rely on volunteers from a single unit (in this case a division) for the MFO-Sinai rotation, and in the case of the MP companies called in support of Operation Maintain Democracy.

Drawing volunteers from a number of units to form a deploying unit affects not only the effectiveness of the deploying unit, but also the readiness of the units from which the volunteers were obtained. The USAR has opposed taking volunteers out of their units for this reason, since the source units suffer an unacceptable degradation in readiness. The USAR favors taking volunteers only from the pool of reservists not assigned to units, the individual ready reserve (IRR), for example. In the case of the MFO-Sinai, however, the USAR was not able to get all of the volunteers needed in the skills required from the IRR pool.

Even where sufficient volunteers have been obtained and have served satisfactorily, there is concern for the future, since it is not clear how large the pool of volunteers may be and whether sufficient individuals with the right skills would be willing to volunteer for future operations. And there is also a concern that even if authority is obtained for involuntary recall, frequent call-ups for OOTW might have a long-term detrimental effect on the readiness of units. Personnel readiness and the ability to retain personnel are uncertain if frequent call-ups persist. Family and employer support may erode if reservists are called too often for operations that are not perceived as of sufficient importance to warrant the disruption of families and businesses.

There is also a lack of information on how long it takes reserve units to recover their readiness for an MRC after participation in an OOTW. The limited data collected by CALL was only for active units.

There is much less experience with extended deployments of reserve units to OOTW and thus much less empirical data to draw on regarding the time needed for a reserve unit to regain readiness after participation in an OOTW. Given that reserve units have a much more limited amount of time to train, however, it should be expected that it will take much longer for a reserve unit to recover from an OOTW deployment, particularly an extended rotation, than for an active like-type unit. In cases where there are sufficient units of a given type in the reserves, this may not be a problem. But for many of the units that are deployed for OOTW this is not the case, and if some of the units are not ready for deployment in the planned time frames, requirements in an MRC will go unfilled for at least some period while units are prepared for deployment.

Readiness degradation will become particularly acute in cases where extended operations require the rotation of reserve units and thus affect a larger fraction of the force. For example, the operation in Bosnia will result in three overseas rotations of some types of reserve units. If the operation were to be extended past the planned date, additional rotations would be required. With even three rotations, there will most likely be three sets of units that will be less ready for deployment to an MRC for some extended period, perhaps years, as a result of participation in the OOTW. For some types of units this would be a problem only in the event of a second MRC. But as discussed earlier in the context of Somalia, for other unit types the Army capability is already taxed for even a limited number of OOTW rotations or individual operations in the event of even one MRC. Quartermaster and transportation capability is marginal at best in these more demanding cases.

IMPLICATIONS FOR THE FUTURE ROLE OF RESERVES IN OOTW

Given today's force structure and policies for utilizing reserve personnel in OOTW, the role of the Army reserves, ARNG and USAR, is not likely to change to any appreciable extent. Deployments to OOTW with a long planning horizon and short rotations, like the historical deployments to SOUTHCOM during annual training periods, are easily accommodated by the reserves. Fast-breaking contingency operations or operations in less benign environments present a much greater challenge for reservists. This is especially true if volunteers must be relied upon to the extent seen in the past.

The use of reserves for future operations at the level experienced recently, and certainly if expanded, will raise continued concerns about the accessibility of reserve component personnel and the availability of units for extended operations. The Army, unlike the Air Force, does not believe it can rely on volunteers for many of its requirements. For some capabilities like civil affairs and psychological operations, most of the Army capability is in the reserves, and reliance on volunteers for these personnel has not been fully successful in the past.

It is somewhat instructive to recall that in the case of both Just Cause and Restore Hope, two CINCs (CINCSOUTH and CINCSOF respectively) planned their operation on the basis of access to the reserve civil affairs and psychological operations personnel through PSRC and requested PSRC from the outset. In neither case were they able to obtain access to other than volunteers. In Just Cause, the reliance on volunteers and the short tour rotations that resulted—in some cases the volunteers could serve only for as little as two weeks to a month—was not nearly as effective as the longer-duration deployment that would have been possible if PSRC authority had been available. Further, PSRC would have permitted the involuntary recall of the specific personnel in civil affairs and psychological operations who had trained for that region and had the particular skills needed. Reliance on volunteers resulted in personnel with other regional orientations and skills and thus a less effective operation.

Expansion of the role of the Army reserves in OOTW also raises the question of costs. Reserve components are not necessarily cheaper than the active if involved in continuous or long-term operations. In the case of the MFO-Sinai, for example, there was an incremental cost to using reserve component personnel for the rotation,[8] perhaps even a greater cost than if an active battalion had been used, as on other rotations. The tradeoff was having an additional active Army infantry battalion ready and available during that period.

---

[8]See GAO (1995b) for an analysis of the historical costs for MFO-Sinai.

There is not money available to pay for additional reserve personnel man-days for participation in OOTW unless it has been planned for in advance. This has been a limiting factor in the use of reservists for some operations (GAO, 1996). Unlike the Air Force, the Army does not include in its budget resources to fund reserve participation in routine peacetime operations. As a result, funds are not available to fund reserve participation in OOTW beyond what can be accommodated during normal annual training periods. Budgeting for routine reserve participation would facilitate the use of reserves in OOTW and permit greater reliance on reserve capabilities.

Even if greater use was made of involuntary recall of reserve forces for OOTW and the Army was able to plan and rely on the reserves for some OOTW, there would remain the issue of the long-term effect on the readiness of the reserve units. This is especially true for personnel readiness, where retention of trained and experienced personnel is key. With frequent OOTW demands, reserve readiness would likely degrade (due to less than full manning) to the point where reserve units would not be able to respond to their primary reason for existence—an MRC or general war.

It should also be pointed out that it is not solely a question of active or reserve personnel from the Army for many of these operations. For much of the support—logistics primarily—there are other alternatives. Use of Department of the Army civilians and commercial contractors, like Brown and Root, has become more prevalent. In recent OOTW in Somalia, Haiti, and Bosnia, civilian contractors have provided many of the required services and replaced the need to deploy additional Army active or reserve personnel. Also, for many of the needed capabilities, one or more of the other services have a like-type capability. Both the Air Force and the Navy have engineering and construction capability, for example, and the Air Force security police squadrons and Marine military police can substitute for Army MPs for some missions. These alternatives lie outside the Army's direct influence, however, and are not as readily effected as Army reserve capabilities.

In addition to budgeting for the planned use of reservists on a more routine basis in peacetime, Army reserves could perhaps play a larger role in future OOTW if the structure was changed somewhat. For example, most active Army support-type units are manned at less than their wartime requirement. Many of the support units are organized at ALO 2 or 3 and, further, are not manned at even these levels due to shortages of personnel in certain skills and/or inefficiencies in the assignment system. As we discussed earlier in the context of Somalia, this results in cross-leveling of personnel into units deploying to OOTW in order to bring them to full strength. An option would be to add sufficient active component personnel to these units in order to bring them to their wartime required strength. This is expensive, however. For some of these units, adding authorizations for individual reserve component personnel to the active unit structure would provide a means, at much less cost, of bringing the unit to full strength for an MRC with personnel who had trained with the unit on an ongoing basis and were familiar with the unit's procedures and personnel.

Adding reservists to the active unit would also provide a pool of personnel to fill in for active personnel deployed to an OOTW and a source of potential volunteers for deployment and participation in the OOTW with elements of the active unit. If such a structure had been available during Operation Restore Democracy in Haiti, for example, reserve military police personnel who had trained with and were familiar with the active unit's local area and peacetime garrison support mission could have been activated (or perhaps participated as volunteers in lieu of drill or annual training) and provided the backfill at Fort Drum, Fort Polk, and Fort Bragg. This would have been a more effective and less disruptive solution than the activation of a unit from California, with volunteers from a number of other units, for service at Fort Drum.

For other types of units it may be better to form sections, platoons, or even companies that would be integrated into active units, but manned by reservists, and would train with the active unit on a continuous basis. If a section or platoon from the active unit were to deploy to an OOTW, the reserve elements could be called in the event of an MRC and be immediately available to deploy with the active elements remaining at home station. A complete unit would then be ready to accomplish its wartime mission, even though some active elements of the unit were still engaged in the OOTW. The first force structure alternative discussed above, using individual reservists, is not unlike that used in the Air Force, where individual drilling IMAs (individual mobilization augmentees) served in active security police squadrons in peacetime and were available to fill out security police air base ground defense flights in the event of war. The second is similar to the Associate Units in the Air Force reserve components. Assignment of reserve component elements, with the required equipment, to active component units would both increase capabilities for OOTW and provide a more robust and flexible active unit for simultaneous use on an OOTW and deployment to an MRC if it should become necessary.

While these force structure alternates have applicability for some units, the most critical and most widely applicable OOTW role for the reserves remains that of augmenting the active component for an MRC or general war. It is here that a greater reliance on the reserve support units may have the greatest potential for mitigating the effects of OOTW on the Army's capability to simultaneously deploy to an MRC while engaged in one or more OOTW. This role would be of value even in those cases where the reserves cannot be used in direct support of the OOTW engagement. When active units are identified for deployment to an OOTW, especially support units that are very limited in number in the active component, it should be recognized that there will be a "hole" in a TPFDD or contingency plan should active combat elements be deployed to any but the smallest of contingencies while the OOTW is ongoing.

As discussed earlier, the active support units that exist in the CONUS constitute the lead elements in the FSP and are fully committed to support the earliest-deploying combat units. When active units from the FSP are deployed to an OOTW, they are not available immediately for an MRC. If the early requirement is to be met, it will have to be met using a reserve unit to deploy more quickly than planned. In many cases there may be a reserve unit that is sufficiently ready that it can be substituted for the active unit without delaying the planned deployment. In most cases, however, steps would need to be taken to upgrade the readiness of the reserve unit to make it available for deployment at the earlier date. Expanding the early-deployment capability of the reserve units in order to substitute for early-deploying active units in the FSP should they be engaged in an OOTW would be more compatible in the long run with the multiple demands on the citizen-soldier than an expanded direct role in OOTW engagement.

Chapter Seven

CONCLUSIONS

Operations other than war are not new to the Army. Over its entire history the Army has participated in disaster relief, humanitarian assistance, response to riots or insurrection, peace enforcement, limited military operations to restore order in foreign countries, refugee resettlement, and other limited military operations short of warfighting. The Army has a long tradition of responding whenever the nation calls, whether to defend its borders, to fight its wars, or simply to assist citizens in their time of need. In addition to being the DoD executive agent for domestic disaster relief, the Army has many capabilities that make it uniquely suited to respond to OOTW requirements both domestic and foreign. It thus finds itself called upon to perform a wide variety of tasks that are not part of its warfighting skill mix and, in some cases, detract from its warfighting readiness.

The nonwarfighting operations that the Army has been called upon to perform have varied widely in size, duration, and type of forces required. In many, the type and size of the force has varied substantially over the course of even a single operation, sometimes changing quickly rather than in any predictable and linear progression from task to task. None have been large in the context of the total Army force, however, and the existing structure has, for the most part, been able to support OOTW requirements while maintaining its readiness and full capability for its primary warfighting mission.

Historically, OOTW requirements have been treated as a lesser included case in force structure planning and in equipping and training the force. Recently, however, two sets of events have occurred that bring into question whether OOTW demands can continue to be so treated without affecting the Army's readiness and capability to perform both OOTW and its primary warfighting mission when called to do so.

First, the demands for OOTW have grown. They have grown both in the sense of more and larger OOTW requirements and in terms of the need to perform the OOTW in perhaps a more efficient and effective manner than ever before. The number of such operations has grown recently, and there is a belief that the number and size may continue to increase. Further, the character and circumstances of many of these operations have placed a greater burden on the Army to perform in the glare of television lights and in view of the world. This has increased the need to be as efficient and as effective as possible and to minimize any chance of either failing in the mission or performing it in a manner that causes national or international political embarrassment or invites condemnation.

Second, these growing demands have occurred while the force structure existing to meet the demands has greatly decreased. This has resulted in fewer units to share in the burden of OOTW and has made it more difficult to maintain sufficient units ready for possible MRCs. There are also fewer training resources to use in training forces for OOTW and retraining them for MRCs after their return. There are fewer dollars in general either to buy specialized equipment and supplies or to maintain the equipment that does exist. These factors force us to ask whether OOTW demands can continue to be treated as a lesser included case, and at what point the Army will either require additional resources to support OOTW or suffer an unacceptable degradation in its MRC capability when it finds itself engaged in OOTW. This is especially relevant in view of the results from our earlier analysis of MRC requirements and Army capabilities, which indicated that although the combat forces are more than adequate for the planned two-MRC case, the support forces are marginal at best.

ARMY FORCES ADEQUATE IN NUMBER FOR MOST UNIT TYPES

Our analysis has shown that for most types of units, the Army has adequate capability to perform both a limited number of OOTW and also maintain its MRC capability. There are a few types of units, however, where the Army force structure is stressed. These units tend to be from the quartermaster and transportation branches but also include some maintenance-type units. They are units that are typically needed in OOTW and are in areas where the Army is short of active units to meet early-deploying MRC requirements even in the absence of an OOTW. Our previous analysis of even the less demanding MRC cases showed that the active component support structure was not adequate to meet some of the desired early-deployment requirements that could not be met by the less-ready reserve component units. The time needed to mobilize and prepare reserve component units means that if the earliest desired capability is to be met, it must be met with active component units.

The analysis also demonstrated that the units whose readiness and availability to deploy to an MRC are affected by an OOTW deployment go beyond the units actually deployed and engaged in the operation. If units are being rotated in the OOTW, many additional units are engaged in preparation and recovery than are currently deployed. Because many OOTW deployments demand different skills and tasks than needed for warfighting, units often require an extended period of preparation and/or training both before and after an OOTW in order to be adequately prepared for it and to regain warfighting skills and readiness after returning from it. Depending upon the tour length policy being followed, the number of units whose readiness and availability are affected (units in preparation or recovery from a rotation) may be two or three times the number actually deployed. Our analysis has taken these effects into account.

As shown in Figure 7.1, there is a shortfall in both quartermaster and transportation capability in the active component units intended to deploy in the earliest phase of an MRC, even in the absence of an OOTW. Adding the requirements for an OOTW like Somalia increases the shortfall, though not to a large degree.

Units like petroleum supply companies, water purification teams, maintenance teams, terminal operations teams and companies, and light-medium and medium truck companies are support units typically deployed to an OOTW and are in short supply in the active component. However, the cases where there are shortfalls caused by OOTW requirements are not large in number, and the cost to com-

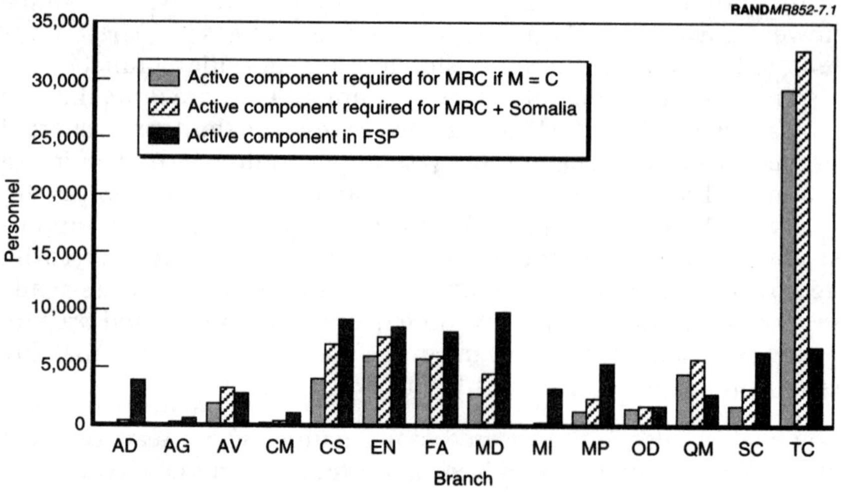

Figure 7.1—Comparison of Requirements and Active Component FSP Capability by Branch

pletely offset the OOTW effects would not be large in terms of the active endstrength required to field the additional units. For example, in the case of Somalia—with only modest reduction in the typical time needed for units to fully regain ready status—the entire affected CSS forces could be replaced with about 7,000 soldiers. For most of the operation the cost would have been only about 4,000–5,000, even if all of the capability needed to be replaced to maintain the MRC capability. For some units there are other alternatives for replacing the active force capability for an MRC.

For example, cargo documentation teams are required in the theater for an MRC by about day fifteen (by about day seven in the case of prepositioned ships). Only three automated cargo documentation units exist in the active component, and two were used in Somalia. However, our analysis of ODS experience would indicate that reserve component units of this type could be relied on to arrive in the theater by about day eighteen. If a two- or three-day delay is acceptable, then the current force would be adequate. If not, it is likely that with only a few additional resources, units like cargo documentation teams or water purification teams could be made ready more quickly and replace active component units engaged in an OOTW and therefore not immediately available for the MRC should it be necessary.

The case is similar for other units like terminal operations teams and perhaps truck units and petroleum supply units that are typically required in OOTW and are in short supply in the active component. These units tend to be single-function support units (in addition, many are rather small teams, as in the case of water purification and cargo documentation) that call for civilian-type skills. These are the type of units the reserve components have demonstrated they are best at providing.[1]

Our analysis has shown that, for the most part, the Army has been able to meet the OOTW demands (as exemplified by operations conducted since 1975) and still have sufficient units available to meet the demands of an MRC. This is not to say that cases cannot be developed where this would not be so. Clearly, if one or more large OOTW involved two Army divisions, as envisioned in the BUR, then virtually all of the active component support force capability would be required, with nothing left in the event of an MRC. Such a case would require PSRC or mobilization so that reserve component units could either be deployed to the OOTW or placed on active duty for training so they could be made ready for early deployment in place of active component units in the event of an MRC. Our results would also not apply if future force reductions were to result in the active component support units being reduced without an offsetting increase in the readiness of reserve component units or some other offsetting actions being taken.

UNIT COUNTS AND ROTATIONS MISS MANY OOTW EFFECTS

Even if units are not being rotated, additional units are affected because personnel are often taken from other units to bring deploying units to full strength. In the case of Somalia, for example, the TPFDD

---

[1]Note also that many of these units require skills that are available in the private sector and could be supplied under contract agreements like those used to obtain similar services from Brown and Root for Somalia, Haiti, and Bosnia.

showed that ten military police companies were deployed to Somalia with a total of 1,193 personnel. Personnel deployment data showed that the military police personnel (MOS CMF 95) that had deployed to Somalia came from 62 different units (41 were military police companies, 10 were military police battalion headquarters units, and 11 were other organizations). This cross-leveling of personnel often occurs because many of the support-type units needed in OOTW are not structured in peacetime with their full wartime-required complement of personnel authorized. Further, because of the drawdown and personnel assignment priorities, these units are also often undermanned even relative to their peacetime authorized level. When called to deploy on a real mission, however, more often than not the unit must be brought to full wartime strength before being deployed.

All of these factors place additional burdens on all the units, whether deployed to the OOTW or not, and require an inordinate amount of command and leadership time and attention to be diverted from normal peacetime training and readiness activities in order to manage the preparation, deployment, sustainment, and recovery of even parts of units deployed to OOTW. Additionally, the Army tailors forces more for OOTW than for MRCs and in nondoctrinal ways, which makes the deployment to an OOTW much more management intensive. So because their commanders and unit leaders are focused on supporting the OOTW, units do not receive sufficient attention to training and readiness for their wartime mission.

Equipment readiness also suffers in many OOTW because of the extra wear and tear caused by intensive usage in often hostile environments and under circumstances that do not permit adequate routine maintenance and service. Support equipment is sometimes lost in transit or given away after the operation, and replacement of that equipment may take a long time. Our analysis, focused on the count of units and their availability, also does not account for any short- or long-term readiness effects caused by the increased stress and personnel tempo resulting from OOTW deployments.

RC FORCES AND OOTW

Like the active component of the Army, the reserve components have a long tradition of OOTW involvement. The National Guard is planned to be the first responder for CONUS OOTW. It is the first force called to assist governors and local authorities and to augment local capabilities. Both components have a long history of use and involvement in OOTW, particularly in SOUTHCOM. They routinely participate in nation building, drug interdiction, and border control operations. But for many OOTW contingencies, the circumstances of the operation and the nature of citizen-soldier service will limit greater direct participation. Nevertheless, there may be options and alternatives for the reserve component to play a greater role in mitigating the effects of the unpredictable nature of OOTW and the day-to-day status of units (active component and reserve component) on the Army's ability to maintain a ready MRC capability.

MITIGATING THE EFFECTS OF OOTW ON MRC CAPABILITY

In many cases, less than full doctrinal Army units are required for OOTW. Forces are tailored to form task forces, particular sections or platoons are required for a specific operation rather than the whole unit, or only a fraction of the capability is needed for the limited operation as compared to the warfighting task. In some cases it makes sense to add active component units or elements—subsistence platoons were added for this reason—to the force structure. In other cases, however, the unpredictable nature and changes in scope of ongoing operations make this a very ineffective use of resources if done to protect against only occasional need. Other alternatives that facilitate the use of reserve component soldiers may be more cost-effective.

In some units, adding a section or platoon to an active component unit manned by part-time reservists would provide the additional capability needed to offset the effects of limited OOTW deployment and permit the early deployment of a fully manned TOE unit in the event of an MRC. In other units it might be necessary to place authorizations for reserve soldiers throughout the unit rather than in sections and platoon formations.

In cases where whole units are typically deployed, it may be better to place greater emphasis on making reserve units more ready so that they might deploy earlier in the event of an MRC. This does not mean keeping some or all of the reserve units more ready all the time. It is likely to be much more cost-effective to establish procedures to support the real-time management of unit and force readiness based on current force needs, problems, and commitments. This would enable the deployment of reserve component units to an MRC earlier than planned when active component units are engaged or committed to engage in an OOTW, while avoiding the expenditure of resources needed to constantly maintain a higher state of readiness even when it is not required by the current situation.

Appendix

FORCE STRUCTURE REQUIREMENT TEMPLATES FOR OOTW

To assess the potential impact of future OOTW on the Army's ability to maintain a ready MRC capability, we first looked at Army involvement in past OOTW. In looking at past OOTW and determining what levels of force involvement future operations might require, we focused on the absolute size of the force for an operation, the branch composition of the force, and characteristics of the operations, e.g., whether units are rotated and the level of combat potential. From this information we constructed a set of force templates to describe in summary the characteristics of the various types of OOTW missions and force requirements.

Each template represents a certain type of OOTW. These operations historically have been very diverse. The templates reflect the wide variety of operations covered by OOTW, including combat operations, humanitarian relief operations, peace mission operations, and operations both within the CONUS and abroad. We found these templates to be useful in exploring how generic types of OOTW might affect Army unit availability and force structure requirements.

The following list of force requirement characteristics are discussed in the OOTW templates:

- the number of personnel simultaneously engaged in the operation,
- the skills of the personnel (or force composition),
- the length of time a unit or individual is deployed,

- the total length of the operation and the need to rotate the personnel or units,
- how time-critical the various parts of the operation are to the complete operation and the length of the planning horizon,
- the risk of combat, and
- the past tendency to use active or reserve component personnel to undertake that type of operation.

OCONUS OPERATIONS

The Army has been involved in a range of OOTW outside the CONUS (OCONUS) since 1975 (USACAA, 1991). Six types of OCONUS operations are summarized in the USACAA report: combat, peacekeeping, foreign nation assistance or nation building, show of force, security augmentation, and humanitarian assistance. We use these six mission types to categorize the templates for OCONUS operations. In addition, we include three other OCONUS operation templates which were not in the USACAA report: peace enforcement missions, which represents missions such as Operation Restore Hope in Somalia; an armed intervention template for missions such as Operation Preserve Democracy in Haiti; and a large armed intervention template representing a possible scenario proposed by the BUR.

Armed Intervention, Combat, and Peace Enforcement Operations

Armed intervention, combat, and peace enforcement operations are intended to coercively compel compliance with international resolutions or patterns of behavior, the primary purpose of which is the maintenance or restoration of peace under conditions broadly acceptable to the international community. The combat forces, and associated support functions, required for these operations will resemble those for major (or lesser) regional conflicts. Threat, terrain, and lack of available infrastructure is liable to increase the role of light and medium forces in these operations as opposed to MRCs (DoD, 1993, and Allen et al., 1993).

Table A.1 indicates the size of forces used in these types of operations and which branch or branches provided the largest contingent.

Table A.1

Armed Intervention, Combat, and Peace Enforcement Operations

Operation	Major Force Component	Indication of Force Size	Force Rotation
Lebanon	Target Acquisition	Co(–)	No
Urgent Fury	IN	Div(–)	No
Just Cause	IN	Div	No
Restore Hope	IN	Div	Yes
Preserve Democracy	IN and MP	Div(–)	Yes
Hypothetical	Inf or Mech	Up to 2 div	Possible

The forces deployed in the combat operations Urgent Fury and Just Cause were of broadly similar size and composition (over 50 percent infantry). Both operations involved fairly short-duration deployments with no requirement to rotate units. They were combat operations, each part was critical to the whole operation, and they involved primarily or only active component units. Thus, the force requirements of these two operations were similar and are likely to be similar to the force requirements of future combat operations. Consequently, Table A.2 includes a combat operation template based upon Just Cause, as this was the larger of the two operations.

The other combat operation in the USACAA (1991) study was the deployment of a target acquisition battery to Lebanon. This involved the deployment of a single unit of fewer than 50 soldiers. A future operation that involved the deployment of a single unit of fewer than 50 soldiers will in general have little impact on the ability of the Army to maintain a ready MRC capability. The most obvious exception is where there is only one unit of that type in the Army and the unit is also required to support an MRC. Similarly, for any type of unit where there are only just enough units to support a single MRC, the deployment of one of them will have an impact on the ability to respond to an MRC. However, such cases should become apparent from the examination of the implication of OOTW with much larger force requirements. Therefore, future combat operations requiring a single small unit are not represented by a template.

Table A.2

Armed Intervention, Combat, and Peace Enforcement Templates

Type of Operation	Force Deployed				Notes
Combat	AD	80	AG	10	10- to 40-day deployment
	AR	40	AV	460	No rotation of force
	CSS	570	EN	230	Risk of combat high
	FA	630	FI	10	Each part critical
	IN	6,340	MAINT	60	Previous operations mainly AC
	MS	130	MI	330	
	MP	1,220	OD	30	
	PA	10	QM	15	
	SC	290	TC	110	
Peace enforcement	Represented by Restore Hope TPFDD				Rotation of units Significant risk of combat Operation mounted quickly Mainly AC, some RC volunteers
Armed intervention	Represented by Preserve Democracy TPFDD				Rotation of units Some risk of combat Operation mounted quickly Previous operations mainly AC
Larger armed intervention	2 divisions plus support troops				Deployment length uncertain Rotation of units uncertain Risk of combat high

Peace enforcement operations may become more common in the future and could take many forms. None of the operations in the USACAA study are of this type. Operation Restore Hope in Somalia and ONUC, the UN operation in the Congo, may represent the only such operations undertaken so far. Operation Restore Hope, as a U.S. operation and much more recent than ONUC, was the basis for the template used in our analysis. The operation was mounted quickly, involved a significant risk of combat (as will any peace enforcement operation), involved the rotation of units, and mainly used active component personnel assisted by a few reserve component volunteers.

In Somalia, the force deployed in later rotations was different from, and smaller than, that initially deployed. This is due to a number of reasons, not least of which is the changing requirement for forces as an operation progresses. In addition, the initial force will include units specifically required to establish the force in theater. In our template we assume that the force required is constant, as the template is intended to represent the composition and number of forces required for the total operation.

There may be a future requirement to undertake operations similar to Preserve Democracy in Haiti. Therefore, the armed intervention template is based upon this operation. The operation was mounted quickly, involved some risk of combat, involved the rotation of units, and used mainly active component personnel assisted by a number of reserve component personnel. In the initial phase of the operation the reserve component personnel were deployed to CONUS locations to replace deployed active component units. Later, special forces units and others were deployed to Haiti to replace active units.

The BUR envisioned the possibility of a larger armed intervention, a lesser regional conflict, and set a level of up to two divisions and support forces (DoD, 1993). Such an intervention would probably involve a high risk of combat. Much less certain is how long the operation might last and whether there would be a need to rotate units. Therefore it would be best to consider two cases. In the first case the force is deployed for a period ranging from a few days to a few months but no unit rotation is required. In the second case the entire force required must be rotated at least once.

Peacekeeping

Peacekeeping operations are noncombat missions, excluding the potential necessity for self-defense, conducted with the consent of all the belligerent parties in order to monitor and facilitate the implementation of cease-fire and peace agreements (Allen et al., 1993). Combat force contributions to peacekeeping operations will in most cases be infantry and SOF intensive and will likely involve force commitments of an extended duration (DoD, 1993). For example, a U.S. Army battalion task force has been part of the Multi-National Force and Observers (MFO) in the Sinai since 1982 (USACAA, 1991).

Peacekeeping operations will typically require heavier concentrations of combat support and combat service support forces than is the case for combat operations. Emphasis will be placed on medical,

Table A.3

Peacekeeping Template

Type of Operation	Force Deployed	Notes
Peacekeeping	IN 550–1,000 CSS 0–360	May continue for years Rotation of units Risk of combat changes Planning horizon changes Initially AC, later mix

engineering, transportation, and command and control facilities (DoD, 1993). For example, the United States provided medical support for the United Nations Protection Force (UNPROFOR) in the former Yugoslavia through a series of six-month rotations (Collins, 1993).

The peacekeeping template is based upon the MFO-Sinai deployment of an infantry battalion. The extended duration of these operations means that units need to be rotated regularly. The initial deployments may need to be undertaken quickly, involve a degree of risk, and tend to be undertaken by the active component. But if the operation continues for a number of years, the degree of risk is likely to decrease significantly, deployments can be planned a long time in the future, and there may be an increase in the use of reserve component personnel, as in the January 1995 MFO-Sinai rotation.

Foreign Nation Assistance or Nation Building

Foreign nation assistance is designed to combat dangers to democratic reform (DoD, 1993), improve allied military capabilities, and facilitate allied combat operations if these become necessary. Such operations are exemplified by the operations conducted by SOUTHCOM in South America. These operations have included the deployment of medical, engineer, military police, military intelligence, public affairs, postal, infantry, artillery, aviation, transportation, logistical, maintenance, special forces, and army band units ranging from two-man cells to battalion-sized elements (ARNG, 1993). Table A.4 provides examples drawn from the USACAA study.

Table A.4

Nation Building Operations

Operation	Major Force Component	Indication of Force Size	Force Rotation
JTF Bravo	MI and AV	Bn/Bde	Yes
Fuertes Caminos	EN	Bn	Yes
Cabanas 85	EN	Bn(+)	No

The USACAA study identified seven nation building and operational support missions, one of which was an amalgamation of various small deployments worldwide between 1986 and 1990. Of the remaining six deployments, all but one, JTF Bravo, were by forces centered on engineer units. The most intensive of these five operations was Cabanas 85, so the "force deployed" section of the template (Table A.5) is based on this operation. However, this operation was different from the others in that the force was not rotated and was drawn from the active component. The other four operations were carried out by the reserve components during annual training periods, with units rotated after two weeks.

This type of operation often continues intermittently over a number of years, involves little risk of combat, and is planned a long time in advance. A more modern example is provided by Operation Fuertes Caminos which, in both 1993 and 1994, involved the deployment of about 5,000 reserve soldiers to Guatemala. It consisted of a series of two-week deployments during annual training, during which a number of engineering and medical tasks were completed (ARNG, 1993).

Table A.5

Nation Building Template

Type of Operation	Force Deployed				Notes
Nation building	CA	5	CSS	100	May continue intermittently for years
	EN	455	MAINT	30	
	MS	55	MP	30	Units deploy for 10–85 days
	QM	120	TC	35	Risk of combat very low
					Long planning horizon
					Both AC and RC used

Show of Force and Security Augmentation

Show of force and security augmentation missions are close enough in size and purpose to be linked together and considered as one. The USACAA study includes seven such operations (see Table A.6). The three operations that involved the deployment of MP units to Panama prior to Just Cause are grouped together as one entry in Table A.7.

Most of the operations in the USACAA study involved the rotation of military police units of up to brigade size. These MP operations often continued for several years. Some deployments were by active component units and others were by reserve component units. In general the risk of combat is probably low, but reserve component MP units, already deployed to Panama when Just Cause occurred, took part in this combat operation. The security augmentation template is based upon the deployment of a military police brigade, as is shown in Table A.7.

The three show of force operations involved the deployment respectively of an aviation battalion, a light infantry brigade, and a mechanized infantry regiment. All of these units were drawn from the active component, possibly due to the potential for these operations to lead to combat and the urgency of some deployments. The shortest operation lasted less than a month, while the longest lasted 21 months and involved the rotation of the unit deployed roughly every four months. The show of force template is based upon these operations, with the force size based on the largest deployment (the mechanized infantry regiment). The possibility of a larger deployment

Table A.6

Show of Force and Security Augmentation Operations

Operation	Major Force Component	Indication of Force Size	Force Rotation
TF Hawk	Aviation	Bn	No
Golden Pheasant	Inf	Bde	No
Nimrod Dancer	Mech Inf	Rgt	No
Philippines	MP	Co	Yes
Panama (3)	MP	Co to Bde	Yes

Table A.7

Show of Force and Security Augmentation Templates

Type of Operation	Force Deployed				Notes
Security augmentation	CA	5	MS	15	May continue for years
	MI	75	MP	970	Units deploy for 10–145 days
	SC	15	TC	50	Risk of combat changes
					Both AC and RC used
Show of force	IN	1,550			May require rotation of units
					Units deploy for 14–222 days
					Can be urgent
					Past operations used AC

exists, as demonstrated by the deployment of the 24th Infantry Division to Kuwait in October 1994. Such larger deployments will have similar force requirements to the armed intervention templates and so are considered to be covered by them.

Humanitarian Assistance

Humanitarian aid and relief operations are similar to domestic natural disaster operations, with the potentially added complications of a long supply chain, language difficulties, different local customs, lack of infrastructure, and local political considerations. Four humanitarian assistance operations are shown in Table A.8.

Operation New Life was the OCONUS part of a large refugee resettlement operation, most of which occurred within the CONUS, therefore it is considered under refugee resettlement below. The Eniwetok cleanup, a three-year effort to remove nuclear-contaminated debris from the Eniwetok Atoll, involved the rotation of various engineer units. This is similar to the requirements of nation building operations and is covered by the nation building template. In other words, a future operation that is similar to the Eniwetok cleanup would place demands on the Army similar to those of a nation building operation as represented by the nation building template. Therefore, the humanitarian assistance template is based upon the other two humanitarian assistance operations, Task Force Crosby and the incident in Guyana. These two operations were concerned

Table A.8

Humanitarian Assistance Operations

Operation	Major Force Component	Indication of Force Size	Force Rotation
Incident in Guyana	CSS	Bn(–)	No
TF Crosby	CSS	Bn(–)	No
Eniwetok Cleanup	EN	Bn(–)	Yes
New Life	IN and CSS	Bde	No

Table A.9

Humanitarian Assistance Template

Type of Operation	Force Deployed				Notes
Humanitarian assistance	CSS	315	MS	15	Short operation up to one month Urgent Negligible risk of combat Past operations used AC

with recovering the remains of U.S. personnel and required similar numbers of soldiers for a similar length of time. They both followed unexpected events, involved a negligible risk of combat, and were undertaken by active component personnel.

More recent operations include Sea Angel in 1991 and Support Hope in 1994. The bulk of the force in Operation Sea Angel, following a cyclone in Bangladesh, was provided by the 5th Marine Expeditionary Brigade, supplemented by an Army contingent of 90 aviation personnel and 15 engineers (McCarthy, 1994). The force deployed to Rwanda in Operation Support Hope was drawn from the Army and Air Force and included combat, combat support, and combat service support personnel.

CONUS OPERATIONS

Domestic operations that arise suddenly, such as natural disasters, civil disturbances, nationwide domestic emergencies, mass immigration, and terrorism, have from time to time overwhelmed the ability of the civil authorities and led to the involvement of Army personnel. Indeed, over half of the OOTW for the fifteen-year period from 1975 to 1990 are domestic operations arising from these causes (USACAA, 1991).

Many of these operations involve only a small number of Army personnel for a brief period, such as the cleanup operation in Hopkinsville, Kentucky, after a tornado in 1978. The impact of such an operation on the Army's ability to maintain a ready MRC capability will be minimal, unless, as discussed above, the Army has only a very few personnel with the required skills. However, these special cases should become apparent from analysis of the operations that require a significant level of Army support, such as the resettlement of Indochinese refugees in 1975.

These operations are domestic operations, so, in general, the National Guard in state status has primary responsibility for providing support when military assistance is required (FM 100-19, 1993). The exceptions are terrorism and nationwide domestic emergencies, such as the 1981 air traffic controller walkout. During fiscal year 1992, the National Guard responded to 322 emergency missions, including four civil disturbances and 112 natural disasters. In more than 97 percent of these cases the National Guard was able to cope without federal assistance (Cook et al., no date). Thus, in most cases, the relevant state's National Guard had sufficient personnel, with the correct skills, available fast enough and for long enough to cope with the emergency.

Disaster Assistance

Disaster assistance is the most common of these OOTW, with 18 of the 27 identified CONUS operations conducted between 1975 and 1990 being for disaster assistance. These operations generally require the removal of debris, preservation of health and safety, and restoration of essential services. The National Guard, as a state organization, has the primary responsibility for providing support when military assistance is required. The role of federal Army forces is to assist civil authorities, under the direction of FEMA, when the magnitude of the disaster exceeds the capabilities of state agencies, including the National Guard (FM 100-19, 1993). Table A.10 provides a number of examples of these operations.

Table A.10

Disaster Assistance Operations

Operation	Major Force Component	Indication of Force Size	Force Rotation
Tornado cleanup, KY	EN	Co(–)	No
Snow removal, OH	EN	Bn	No
1989 CA earthquake	EN and IN	Bde	No
Hurricane Hugo, SC	EN	Bde	No

Most disaster assistance operations are small; 12 of the 18 identified operations involved fewer than 1,000 soldiers (both active and reserve) for only a few days. The largest number of soldiers involved in a disaster assistance operation was 3,608 on a fire-fighting mission in Yellowstone in 1989. The number of soldiers required for disaster assistance can be significantly higher; in 1992, following Hurricane Andrew, 25,000 active Army personnel, together with 6,000 National Guardsmen, were required for the disaster assistance operation in Florida (Lynch, 1993). Therefore, there are two disaster assistance templates, identical except for the size of the force required (and exact skill mix). The small template is based upon the 1989 California earthquake, while the larger is based upon Hurricane Andrew.

Table A.11

Disaster Assistance Templates

Type of Operation	Force Deployed				Notes
Disaster assistance (moderate)	AG	520	AV	10	Operation up to one month long
	CSS	150	EN	1,025	Urgent
	IN	925	MS	130	Mix of AC and RC
Disaster assistance (large)	25,000 AC and 6,000 RC				

Civil Disturbances

Civil disturbances may range from unruly demonstrations to widespread rioting with looting and arson. The National Guard, as a state organization, is likely to provide the first military response to any civil disturbance. Federal Army forces may assist local and state civil authorities under certain circumstances. The primary function of Army personnel during civil disturbance missions is to patrol the area and provide a visible presence, with the aim of preventing unlawful acts. They may also be used to disperse unlawful assemblies, distribute essential goods, maintain essential services, establish traffic control points, and serve as quick-reaction forces (FM 100-19, 1993 and Lynch, 1993). Three civil disturbance operations are shown in Table A.12, including the 1987 prison disturbances in Oakdale, Louisiana, and Atlanta, Georgia, Operation Hawkeye in the Virgin Islands following Hurricane Hugo, and the Los Angeles riots in 1992.

These operations, like disaster assistance operations, are often small. For example, the prison disturbances during 1987 involved just 500 Army personnel. However, some operations can require a significant number of Army personnel. For example, a total of 6,000 National Guardsmen and 4,500 active soldiers were deployed in Los Angeles in 1992 to help state and local law enforcement officials control the riots (Lynch, 1993). Operation Hawkeye was a mid-sized operation and is used to size the law enforcement template. A larger operation, such as that during the Los Angeles riots, can be considered as equivalent to undertaking a number of operations simultaneously or a scaling of the forces shown.

Table A.12

Civil Disturbance Operations

Operation	Major Force Component	Indication of Force Size	Force Rotation
Hawkeye	MP	Bde	No
Prison disturbance	MS, MP, EN	Bn	No
LA riots		Div(–)	No

Table A.13

Law Enforcement Support Template

Type of Operation	Force Deployed				Notes
Law enforcement support	AV	25	CAV	10	Operation up to 1.5 months
	CSS	90	EN	45	Urgent
	MS	55	MI	20	Mix of AC and RC
	MP	1,110	PA	5	
	SC	25	CA & PSYOP	45	

Counterterrorism

The Federal Bureau of Investigation is the lead law enforcement agency concerning incidents of terrorism in the United States. The Army may provide counterterrorism assistance, which is essentially similar to that provided in civil disturbances. The Army also undertakes antiterrorism actions aimed at reducing the vulnerability of Army personnel and property to terrorist attacks. Army Explosive Ordnance Disposal (EOD) personnel support the U.S. Secret Service and may assist local law enforcement agencies (FM 100-19, 1993). The present requirement for Army support to combat terrorism is small. The USACAA study provides only one example of such an operation, which involved the deployment of two MP companies to reinforce Seneca Army Depot for almost four months in 1983. These operations have a similar force requirement to that for law enforcement, so future operations of similar type are covered by the law enforcement template in Table A.13. If a future terrorist threat required a much longer-running operation, then it would have force requirements similar to those in the security augmentation template in Table A.7. Therefore we have not developed a specific counterterrorism template.

Nationwide Domestic Emergencies

The Army has provided assistance to civil authorities during nationwide domestic emergencies, such as the 1970 postal strike and the 1981 air traffic controller walkout. These operations are infrequent, may require specialist skills, and can last a significant length of time.

However, they are sufficiently infrequent and likely to be of such a specialized nature that they are not represented by a template.

Refugee Resettlement

The Army has provided assistance during refugee resettlement operations that arise as the result of mass immigrations. Between 1975 and 1990, domestic refugee resettlement operations accounted for 66.9 percent of the man-days used in domestic OOTW, though only 5 of the 27 operations were of this type (USACAA, 1991). Hence, these operations are often large. The figures are even more striking when it is recognized that three of these operations were part of the 1980 Cuban resettlement and the other two were part of the 1975 Indochinese resettlement. Table A.14 provides an outline of the forces required for these five operations. In addition, the largest of the OCONUS humanitarian assistance operations, New Life, was part of the 1975 Indochinese resettlement (see Table A.8).

In total, the 1975 Indochinese resettlement required almost 6,000 soldiers, nearly all from the active component, for about 200 days. The 1980 Cuban resettlement was of similar size but involved a much higher number of reserve component soldiers. The refugee resettlement template is based upon these operations.

Both Cuban and Haitian refugees were housed at Guantanamo Naval Base in 1994 by Joint Task Force 160. In August 1994, Joint Task Force 160 consisted of over 1,200 Army soldiers, with the largest single contingent being about 500 MPs. The Task Force was planned to expand to perhaps 2,400 personnel as the number of Cuban refugees increased rapidly in late August (Naylor, 1994). The 1994 operation was smaller than the 1980 and 1975 operations, but future operations could be as large.

Table A.14

Refugee Resettlement Operations

Operation	Major Force Component	Indication of Force Size	Force Rotation
Indochinese (2)	MP, EN, and CSS	Bde(+)	No
Cuban 1980/81 (3)	MP	Bde(+)	No

Table A.15

Refugee Resettlement Template

Type of Operation	Force Deployed				Notes
Refugee resettlement	AG	370	CA	115	Operation 6–8 months long
	CSS	1,340	EN	755	Urgent
	FA	100	IN	1,080	Mix of AC and RC
	MS	680	MP	820	
	MI	5	OD	130	
	PI	20	SC	20	
	PSYOP	80	TC	235	
	MAINT	125			

Other Domestic Operations

Other domestic operations to which the Army contributes are domestic community assistance and the counterdrug campaign. Domestic community assistance operations fulfill community needs that would not otherwise be met and involve a wide range of activities, including public works maintenance and management, civilian youth opportunities program, medical readiness program, national events, community liaison, physical improvements and social improvements.

These operations may take advantage of the general military skills and capabilities found in active or reserve units. For example, a program started in 1992 by Fort Eustis, called Operation Self-Enhancement, gives high-risk middle school students the opportunity to visit the post, where they receive light military training provided by members of a cadre team (FM 100-19, 1993). Alternatively, these operations depend upon the specific skills, usually medical or engineering, of an active or reserve unit. One such case was the deployment of about 75 personnel from the 147th Combat Support Hospital to a blighted Denver neighborhood for two weeks to help with local health care in March 1994 (*Washington Times,* 1993).

The counterdrug campaign has Army units providing support to law enforcement agencies. This support includes training, operational support, intelligence, detection, reconnaissance, transportation, and facilities. The *Posse Comitatus* Act applies to federal Army forces, but not to the National Guard in state status, so this act makes it easier to use the National Guard, under state control, to support counterdrug operations rather than the active Army or Army Reserve (FM 100-19, 1993).

The number of man-days of effort assigned to domestic community assistance and the counterdrug campaign is significant. For example, in fiscal year 1992 the National Guard took part in 5,926 counterdrug operations involving a total of 1,092,319 man-days (Posehn, 1993). However, the number of personnel involved on a particular operation under either of these umbrella terms is small. In addition, the personnel engaged in these tasks would be available to respond quickly to an MRC, so these operations will have no impact on the ability of the Army to respond to an MRC. Consequently, we did not create templates for these operations.

REFERENCES

Allen, W. W., A. D. Johnson, and J. T. Nelsen II, "Peacekeeping and Peace Enforcement Operations," *Military Review,* Vol. 73, No. 10, October 1993, pp. 53–61.

Army National Guard, *FY 94 ARNG Central American and Southern American Overview,* NGB-ARO-OY, Washington, D.C., August 9, 1993.

Army Times, "Griffith: 'Endeavor' Hampers Two-war Strategy," Washington, D.C., May 6, 1996, p. 3.

——— "Guard Brings Promise to Albania," Washington, D.C., April 15, 1996, pp. 16.

Brinkerhoff, John R., *Case Studies in Reserve Component Volunteerism: The 670th Military Police Company in Operations UPHOLD DEMOCRACY,* Alexandria, VA: Institute for Defense Analyses, IDA Document D-1663, May 1995a.

———, *Case Studies in Reserve Component Volunteerism: A Composite Battalion Task Force for the U.S. Army Element of the Multinational Force and Observers Mission, Sinai,* Alexandria, VA: Institute for Defense Analyses, IDA Document D-1665, May 1995b.

———, *Case Studies in Reserve Component Volunteerism: The 258th Quartermaster Supply Company,* Alexandria, VA: Institute for Defense Analyses, IDA Document D-1668, May 1995c.

Brown, Roger Allen, William Fedorochko, Jr., and John F. Schank, *Assessing the State and Federal Missions of the National Guard,* Santa Monica, CA: RAND, MR-557-A, 1995.

Brown, Roger A., John F. Schank, Carl Dahlman, and Leslie Lewis, *Assessing the Potential for Using Reserves in Operations Other Than War,* Santa Monica, CA: RAND, MR-796-OSD, 1997.

Collins, John, *Roles and Functions of U.S. Combat Forces, Past, Present and Prospects,* Washington, D.C.: CRS Report for Congress, January 21, 1993.

Cook III, LTC Frank J., LTC Jeffrey A. McChesney, LTC Gary R. Stephens, LTC Gregory G. Wilmoth, and COL William M. Wilson, *The Defense Department's Role in Humanitarian and Disaster Relief,* Harvard National Security Fellows, Final Draft, no date.

Davis, Lois M., Susan Hosek, Michael Tate, Mark Perry, Gerard Hepler, and Paul S. Steinberg, *Army Medical Support for Operations Other Than War,* Santa Monica, CA: RAND, MR-773-A, 1997.

Defense Week, "Outsourcing Bosnia Base Work Saved Millions, Analysis Shows," July 15, 1996, p. 1.

Department of Defense (DoD), *Report On the Bottom-Up Review,* Washington, D.C., October 1993.

———, *Accessibility of Reserve Component Forces,* Washington, D.C., April 18, 1994.

Diamond, John, "Future War No Place for Short-Timers," *San Diego Union-Tribune,* March 6, 1996, p. 5.

General Accounting Office (GAO), *Peace Operations: Heavy Use of Key Capabilities May Affect Response to Regional Conflicts,* Washington, D.C.: GAO/NSIAD-95-51, March 8, 1995a.

———, *Peacekeeping: Assessment of U.S. Participation in the Multinational Force and Observers,* Washington, D.C.: GAO/NSIAD-95-113, August 15, 1995b.

———, *Peace Operation: Effect of Training, Equipment, and Other Factors on Unit Capability,* Washington, D.C.: GAO/NSIAD-96-14, October 18, 1995c.

———, *Peace Operation: Reservists Have Volunteered When Needed,* Washington, D.C.: GAO/NSIAD-96-75, April 26, 1996.

Hartford Courant, "Troop's Mission: Healing," March 7, 1994, p. C1.

Headquarters, Department of the Army (HqDA), *Army Focus,* Washington, D.C., September 1993.

———, *DECISIVE VICTORY America's Power Projection Army,* Washington, D.C., October 1994a.

———, *USA Posture Statement FY95 Challenges & Opportunities,* Washington, D.C., February 1994b.

———, FM 100-5, *Operations,* Washington, D.C., 14 June 1993.

———, U.S. Marine Corps, FM 100-19, FM 7-10, *Domestic Support Operations,* Washington, D.C., July 1993.

Lippiatt, Thomas F., J. Michael Polich, and Ronald E. Sortor, *Post-Mobilization Training of Army Reserve Component Combat Units,* Santa Monica, CA: RAND, MR-124-A, 1992.

Lippiatt, Thomas F., J. Michael Polich, Ronald E. Sortor, and Patricia K. Dey, *Mobilization and Train-Up Times for Army Reserve Component Support Units,* Santa Monica, CA: RAND, MR-125-A, 1992.

Lynch, David J., "When the Mission Is Aid," *Air Force Magazine,* February 1993.

McCarthy, Paul A., *Operation Sea Angel: A Case Study,* Santa Monica, CA: RAND, MR-374-A, 1994.

Naylor, Sean D., "U.S. May Double Force at Guantanamo Bay," *Army Times,* September 5, 1994, p. 22.

Phelps, Ruth H., and Beatrice J. Farr, *Reserve Component Soldiers as Peacekeepers,* Washington, D.C.: U.S. Army Research Institute for Behavioral Sciences (ARI), September 1996.

Pinch, Franklin C., *Lessons from Canadian Peacekeeping Experiences: A Human Resource Perspective,* Army Research Institute, August 1994.

Posehn, LTC Norris C., *National Guard Counter Drug Support, Supply and Demand Reduction Activities Summary,* unpublished briefing slides, El Paso, TX, January 1994.

Schrader, John Y., *The Army's Role in Domestic Disaster Support: An Assessment of Policy Choices,* RAND, MR-303-A, Santa Monica, 1993.

Shelton, Henry, "Impact of Operations Other Than War," Fort Bragg, NC, February 22, 1995.

Sortor, Ronald E., *Army Active/Reserve Mix: Force Planning for Major Regional Contingencies,* Santa Monica, CA: RAND, MR-545-A, 1995.

Sortor, Ronald E., Thomas F. Lippiatt, and J. Michael Polich, *Planning Reserve Mobilization: Inferences from Operation Desert Shield,* Santa Monica, CA: RAND, MR-123-A, 1993.

Sortor, Ronald E., Thomas F. Lippiatt, J. Michael Polich, and James C. Crowley, *Training Readiness in the Army Reserve Components,* Santa Monica, CA: RAND, MR-474-A, 1994.

Taw, Jennifer M., *Operation Just Cause: Lessons for Operations Other Than War,* Santa Monica, CA: RAND, MR-569-A, 1996.

Taw, Jennifer M., and John E. Peters, *Operations Other Than War: Implications for the U.S. Army,* RAND, Santa Monica, CA: MR-566-A, 1995.

U.S. Army Center for Army Lessons Learned, *The Effects of Peace Operations on Unit Readiness, Interim Report,* Fort Leavenworth, KS, June 1995.

U.S. Army Concepts Analysis Agency (USACAA), *Force Employment Study (FES),* Bethesda, MD: CAA-SR-91-4, February 1991.

U.S. Army Training and Doctrine Command (TRADOC), *TRADOC Pamphlet 525-5: Force XXI Operations,* Fort Monroe, August 1994.

Washington Times, "Guard Sets Up Tent Clinic for Needy," March 10, 1994, p. 3.